ASPATORE
BOOKS

About Aspatore Books
The Biggest Names in Business Books & More....
www.Aspatore.com

Aspatore Books has become one of the leading business book publishers in record setting time by publishing an all-star cast of C-level (CEO, CTO, CFO, COO, CMO) leaders from over half the Fortune 100 companies and other leading executives. Aspatore Books publishes the Inside the Minds, Bigwig Briefs, ExecEnablers and Aspatore Business Review imprints in addition to other best selling non-fiction books. Aspatore Books is focused on infusing creativity, innovation and interaction into the book publishing industry and providing a heightened experience for readers worldwide. Aspatore Books focuses on publishing traditional print books, while our portfolio company, Big Brand Books focuses on developing areas within the book-publishing world. Aspatore Books is committed to providing our readers, authors, bookstores, distributors and customers with the highest quality books, book related services, and publishing execution available anywhere in the world.

About Big Brand Books

Big Brand Books assists leading companies and select individuals with book writing, publisher negotiations, book publishing, book sponsorship, worldwide book promotion and generating a new revenue stream from publishing. Services also include white paper, briefing, research report, bulletin, newsletter and article writing, editing, marketing and distribution. The goal of Big Brand Books is to help our clients capture the attention of prospective customers, retain loyal clients and penetrate new target markets by sharing valuable information in publications and providing the highest quality content for readers worldwide. For more information please visit www.BigBrandBooks.com or email jonp@bigbrandbooks.com.

INSIDE THE MINDS

Inside the Minds:
Chief Technology Officers

Industry Experts Reveal the Secrets to Developing, Implementing, and Capitalizing on the Best Technologies in the World

ASPATORE BOOKS

Published by Aspatore Books, Inc.
For information on bulk orders, sponsorship opportunities or any other questions please email store@aspatore.com. For corrections, company/title updates, comments or any other inquiries please email info@aspatore.com.

First Printing, December 2000
10 9 8 7 6 5 4 3 2 1

ISBN 1-58762-008-1

Library of Congress Card Number: 00-111059

Cover design by Michael Lepera/Ariosto Graphics & James Weinberg

Material in this book is for educational purposes only. This book is sold with the understanding that neither any of the authors or the publisher is engaged in rendering legal, accounting, investment, or any other professional service.

This book is printed on acid free paper.

A special thanks to Brigeth Rivera, Ted Juliano, Kirsten Catanzano, Susan Chernauskas, and Melissa Conradi

The views expressed by the individuals in this book do not necessarily reflect the views shared by the companies they are employed by (or the companies mentioned in this book). The companies referenced may not be the same company that the individual works for since the publishing of this book.

About Inside the Minds
Business Intelligence From Industry Insiders

Become a Part of
Inside the Minds

Suggest a Topic For an Upcoming Book, Become a Reviewer, Nominate an Executive, Post Comments on the Topics Mentioned, Read Expanded Excerpts, Free Excerpts From Upcoming Books

www.InsideTheMinds.com

Inside the Minds was conceived in order to give readers actual insights into the leading minds of business executives worldwide. Because so few books or other publications are actually written by executives in industry, *Inside the Minds* presents an unprecedented look at various industries and professions never before available. Each chapter is comparable to a white paper and is a very future oriented look at where their industry/profession is heading. In addition, the *Inside the Minds* web site makes the reading experience interactive by enabling readers to post messages and interact with each other, become a reviewer for upcoming books, read expanded comments on the topics covered and nominate individuals for upcoming books. The *Inside the Minds* series is revolutionizing the business book market by publishing an unparalleled group of executives and providing an unprecedented introspective look into the leading minds of the business world.

ASPATORE TECHNOLOGY REVIEW
Tear Out This Page and Mail or Fax To:

Aspatore Books, PO Box 883, Bedford, MA 01730
Or Fax To (617) 249-1970

Name:

Email:

Shipping Address:

City: State: Zip:

Billing Address:

City: State: Zip:

Phone:

Lock in at the Current Rates Today-Rates Increase Every Year
Please Check the Desired Length Subscription:

1 Year ($1,090) _____   2 Years (Save 10%-$1,962) _____
5 Years (Save 20%-$4,360) _____   10 Years (Save 30%-$7,630) _____
Lifetime Subscription ($24,980) _____

(If mailing in a check you can skip this section but please read fine print below and sign below)
Credit Card Type (Visa & Mastercard & Amex):

Credit Card Number:

Expiration Date:

Signature:

Would you like us to automatically bill your credit card at the end of your
subscription so there is no discontinuity in service? (You can still cancel your
subscription at any point before the renewal date.) Please circle: Yes   No

***(Please note the billing address much match the address on file with your credit
card company exactly)**

Terms & Conditions
We shall send a confirmation receipt to your email address. If ordering from Massachusetts, please add
5% sales tax on the order (not including shipping and handling). If ordering from outside of the US, an
additional $51.95 per year will be charged for shipping and handling costs. All issues are paperback and
will be shipped as soon as they become available. Sorry, no returns or refunds at any point unless
automatic billing is selected, at which point you may cancel at any time before your subscription is
renewed (no funds shall be returned however for the period currently subscribed to). Issues that are not
already published will be shipped upon publication date. Publication dates are subject to delay-please
allow 1-2 weeks for delivery of first issue. If a new issue is not coming out for another month, the issue
from the previous quarter will be sent for the first issue. For the most up to date information on
publication dates and availability please visit www.Aspatore.com.

Inside the Minds:
Chief Technology Officers

Industry Experts Reveal the Secrets to Developing, Implementing, and Capitalizing on the Best Technologies in the World

CONTENTS

Andrew Wolfe, SONICblue (Formerly S3)
Staying on Top of Changing Technologies   11

Neil Webber, Vignette
Building What the Market Needs   41

Dwight Gibbs, The Motley Fool
Let the Business Dictate the Technology   59

Peter Stern, Datek
Building Your Technology Solution From the Ground Up   83

Warwick Ford, VeriSign
The Security Behind Technologies   111

Ron Moritz, Symantec
Building Leading Technology   129

Dermot McCormack, Flooz.com
The Business Sense Behind Technology   157

Pavan Nigam, WebMD
Building a Simple & Scalable Technology Interface   185

Michael Wolfe, Kana Communications
Designing the Right Technology Solution   207

Daniel Jaye, Engage
The Role of a CTO   229

ANDREW WOLFE
Staying on Top of Changing Technologies
SONICblue (Formerly S3)
CTO

Backgound

I joined S3 about three and a half years ago. Prior to that I had been an electrical engineering professor at Princeton University, where I did teaching and research in media processor architecture and embedded computing. Prior to that I was a graduate student and I had a startup of my own, where we developed touch screens and sensors for interactive computing for kiosks and some of the first PDA type devices.

The CTO Position

Obviously it's very different at every company, but at S3 my primary responsibility is to help set the strategic direction of the company as a whole and to develop and maintain relationships with various kinds of technology partners who will help us to build better products in the future. Also, I work with the various product groups in terms of product planning, and I help them determine how we can bring new features and new capabilities to our various products, and how to get our products to work together in more interesting ways so that we can come up with more of a total solution than a group of individual products.

How the CTO Role Varies From Company to Company

The biggest difference is that this is a product company. For a lot of Internet companies a CTO is somewhat of an information systems manager, where they set up the infrastructure to provide information services. They design the architecture of server farms and software systems and what not.

S3 is not an information services company, it's a product company. It's more like a consumer electronics company. So my role is not in information systems, but rather in helping our divisions design products that our customers will later use to connect to various information systems.

Qualities a CTO Needs to be Successful

I think a CTO needs to have a very broad knowledge of the industry to be able to understand new technologies very quickly, because they change every day. It's also important to be able to develop and maintain relationships with a wide variety of other kinds of partners in the industry. Being a CTO is primarily a listening job: Listening to what's going on at other companies, listening to what's going on with your competitors, listening to what's going on in startups that are developing new markets and new product categories, and listening to customers and understanding what kind of problems they're trying to solve and what kind of new technologies they want to see in the future. CTOs are taking all that knowledge, all those things you can go out and glean from other people, condensing it, filtering it, and spreading that knowledge throughout the company so that people here can develop products that address a lot of these outside concerns.

How the CTO Interacts With the CEO

The CEO calls me about every 20 minutes and I come to his office. This is the only place I've ever been a CTO, but here it's a very, very close relationship. We talk several times a day. He confers with me on a wide range of strategic issues. I consult him every time there's a decision that needs to be made that really effects the direction of the company. Basically, he has sent me on a mission to go out and find opportunities for the company to get into new areas, new technologies, and new relationships. Then I go out and find those opportunities and filter them. We see maybe 300-400 opportunities a year to work with other companies.

I assess them, make recommendations, and then take them back to him, where he essentially makes the final call as to where the dollars go.

Building a Technology Team

The most important thing to look for is smart people. There's no question about it. A few very smart, creative people go a long way in terms of making a company successful. There's nothing that can make up for that.

The other thing you need is a vision. You need a bunch of people who can work together who know very specifically how they want to somehow change the world in the next year or two or sometimes a little bit longer timeframe depending on the industry. But for the types of things we're doing that are very Internet or network product related, our timeframe is a year or two. We are always asking, how do we change what people do; how do we change how they do it? And if you can get a group of people all thinking on the same page in terms of accomplishing something great and something new, you'll have a good team. In our case the big idea lately has been to change the way people listen to music. For the last year or two that's been our big push with the RIO product line.

If you can get a group of smart, preferably experienced people to start thinking about how you can change the way that consumers do something, they very quickly start to generate new ideas and new solutions, and then it's easy to get a lot of other people excited about working with them to deliver a product.

Dominating Technologies in the Future

The next five years is a little bit easier for me, because we've focused our attention on the huge change that's caused by ubiquitous access to the Internet.

Over the next several years, access to information, to data, is going to become a utility just like we think of electricity or water today. We know that we can design a huge range of products today that can depend on electricity for their operation, and it can do all kinds of cool things. It can light up our house, refrigerate our food, all kinds of stuff. Information is going to start to become that same kind of ubiquitous utility, where I can create all kinds of devices that I'm going to use at work or at home that have access to information that I've stored locally on local servers, information that belongs to other people that's out on the Internet someplace, information that's created in real time, and that information's going to be very, very broad in scope. Some of it's going to be music, some of it's going to be video, some of it's going to be information stored in databases, some of it's going to be news or stock quotes or other kinds of things like that. We've already hit the point where today sitting down at my computer I can get a lot of those things. I can get almost all of those things. But new kinds of networking—broadband, Internet connectivity, and broadband home networking including both new wired technologies and new wireless technologies—are going to make it so that I can afford to put that kind of connectivity into hundreds of different kinds of devices, so I can have devices that I can use in lots and lots of different places.

I'm going to have various kinds of devices in all the different rooms of my house where I have access to information around the world. That's going to change the way that I spend my day. It's going to change the way I listen to music. It's going to change the way I watch TV. It's going to change the way I get news in the morning. It's going to change the way I spend my money. And that creates huge new opportunities for new kinds of products, new kinds of business models, and changes in the way that businesses interact with consumers.

Keeping on Top of Technologies

Well, talking to smart people is the best way. The important part of being a CTO is having a good network of contacts, talking to people here in Silicon Valley. It's one of the reasons why Silicon Valley is a great place to be: Every week I can have lunch with somebody who knows part of the industry that I don't, who can tell me what's going on, where to look, and what's happening.

It really helps when a company is eager and willing to invest in new technologies, as S3 is, because people with new ideas come to you. That's another good way to learn about changes and new ideas. And through my background

I still have a lot of connections in the university community. A lot of college professors in engineering and in computer science can help keep me up to speed on a lot of what's going on. There's also just a lot of reading involved. Keeping up with what's in the news, what's in the trade media, talking to suppliers about what they're doing, talking to customers about what kinds of new opportunities they see, and then kind of putting all that information together.

Technologies Strategic to the Organization

Home networking and broadband connectivity are really the two biggest right now. What that does is allow things to talk to each other. When things can talk to each other they can do more interesting things than they can when they're standing alone.

The other thing that goes along with connectivity is media data compression. Compression makes entertainment and advanced communication over the Internet practical. The obvious influences are MP3 for audio and Mpeg2 for video, but there are a lot of new technologies around as well. There's Microsoft WMA, and Dolby AAC, and a bunch of different compression technologies. Which technology you use is far less important than the fact that we can now move entertainment information around digitally very effectively, and that allows us to connect all of our entertainment devices, to obtain entertainment over the Internet and to build new kinds of entertainment form factors. Things can be much smaller, much cheaper, and more portable than they were before. That's letting us start to do really cool new things, initially with music and later on with video.

Evolving Technologies

In addition to the ones I've talked about, wireless technologies are key. Wireless local area networking is going to enable people to have new kinds of portable devices in the home like information Web pads. We're developing those through a subsidiary called FrontPath. We'll see new products like portable music devices that directly connect over wireless from a person's car to a server that is inside their house. Wide area wireless will also be important. In other words, third-generation cell phone technology. Things that will allow high bandwidth wireless data connectivity when you're not in your home. I think those are the two biggest long-term strategic technologies for us.

Dealing With Change in the Industry

Change is interesting because there are two kinds of change in the industry. There's predictable change and there's unpredictable change. At S3, our background is as a semiconductor company and we have a great understanding of semiconductors. There are areas like semiconductor integration, semiconductor clock speeds, and network bandwidth where we can draw a very predictable curve and we can predict with amazing accuracy how these things are going to change over the years. We know we're going to be able to build the same capabilities onto a chip for half as much money 18 months from now as we are today. We know that we're going to get 100 megabytes per second broadband connectivity five to ten years from now. Those things are pretty predictable, and being able to model those things, to plan products and product strategies around them, and to work on some of the more complex issues like product definition, infrastructure development, and software technology based on those assumptions allows you to build your business model around predicting those changes in technology.

The harder problem comes up in places where you can't predict what's going to happen. For example, one that's been very difficult for us has been HDTV. Technically we know how to do lots of interesting things with digital television, but the market is so complicated that we don't know when it's going to be a viable commercial mass market technology.

We know there are going to be transitions in cell phones to 3G technology. It's very hard for us to predict when that's going to happen.

We know that people will make a move from physical distribution of music and video to Internet distribution of music and video, but there are lots of complex issues in terms of security and business models that are very difficult for us to predict.

So those are the much tougher issues to deal with than core technology, and we deal with those by trying to be prepared for multiple situations, trying to be able to demonstrate new technologies early so that we can encourage people to make changes. In some cases trying to jump out ahead and control changes.

For example, some of the emerging home networking standards are very viable in the United States but they have some regulatory issues in Europe. So we put together a partnership and a consortium with 13 other major electronics companies where we're trying to persuade European governments to make some changes in the regulations to allow us to use the same technology in the U.S. and Europe.

So there are some things that we can do to try to control changes in technology, but in a lot of cases we just have to be very nimble and continue to ask questions. We have to continue to check the pulse of the user community and our customers on almost a monthly basis to try to understand how things are changing and try to react very quickly.

Keeping the Organization Technologically Nimble

You keep small teams working on projects, you share information very effectively between teams, you make sure that many, many people within the organization have contact with customers and suppliers so everybody's getting a lot of external input. You reward people based on internal measures of success, not necessarily just on whether or not a product is successful in the market. You want to encourage people to work on new technologies, even if there's some doubt as to whether or not they're going to be successful in the market, because you want to be ready and you want to have multiple approaches to solving any one problem. And you move people from problem area to problem area so that they gain experience, they gain breadth, and they learn how to start working on new problems very, very quickly.

The Technology Advantage

We're in a number of different markets, and it differs in different ones. Let's focus on our strongest growth market right now, which is the digital music space. In the digital music space we've been able to integrate many different music technologies, including decoding, security, and interoperability with PCs. Take that together with some very aggressive design skills. Good industrial design, good packaging, good power management, and some basic engineering discipline and simply put those together and create a platform that solves more problems for a customer than anyone else can this year. And in order to keep that lead, we have very tight contact with the music industry, we have very tight contact with silicon suppliers, and with content suppliers and with software suppliers, and we have a large stream of people that goes out and finds the technologies that are going to be interesting in the next 12 to 18 months, and rapidly integrates them into our product platform. Our platform is very modular, very expandable, and it's designed so that as new features become interesting we can get those new features into the product and out on the market faster than anyone else.

And in the consumer product business, where you don't want to have products or features that are completely different from everyone else because you want to be based around standards, you want to have interoperability. Having a product architecture and product design teams that can bring features to the market quickly is probably your strongest advantage.

Resources in Demand

Technical skills are very much in demand. Skilled marketing people are always in demand. In the end a lot of this business is about finding out what people want, finding out what people will pay for, and defining it correctly the first time. These are all very schedule-driven businesses. You have to get things right. Having marketing teams that can go out and understand consumer needs and then position your product to solve those consumer needs is critical as cycles get shorter.

The other thing that's very difficult is program management. We tend to focus very heavily on the creative part of the engineering—the guys who sit down and write the software for a new feature, or who design the user interface, or who design the circuits for a new product. That's all critical; we couldn't operate without that. But there's another piece that really defines successful companies, and that is the ability to do things right the first time, the ability to do things on time, on budget, and the program management skills, the people who can manage projects, keep them on schedule, understand the technology, understand the risks and manage those risks, are critical, and they are very, very difficult to find.

Strategy for Acquiring People

You go find other successful companies and you try to convince their people that your product is more interesting so they will come work for you.

In terms of management resources, experience is key. In most cases you just try to go find experienced people out there in the industry and attract them to your company by being in a more exciting product area.

In some cases we've also found people within our own company with engineering jobs who seem to have the skills to do this well, who have been through several projects and are ready to simply move to a different role, and we've done that as well.

The Future in Terms of Platforms and Programming Languages

In some cases we're real jump-on-the-bandwagon types. In other cases we aren't. It really depends on what our products do.

We look at everything thinking, what does this product need and how are we going to get there? In some cases, going with the hot, new technology makes sense because there is lots of development around it, there are a lot of people interested in it, there are a lot of people who are up to speed on the technology and want to work with it. We can attract talent that way.

In other cases, it doesn't make any sense for us, and let me give you some examples.

We've found Linux to be a great platform for us. It's a solid operating system. As we find that products need new features, we can often go out to the Linux community and find that they've already been developed and that we can simply immediately incorporate them in.

So, for example, as we evolved our Rio products from hand-held music devices to networked music systems, by basing many of those networked music devices on Linux, much of the networking software was freely available and could immediately be incorporated into the product.

It's also very easy for us to go out and find experienced Linux developers, so we've been able to build teams very quickly. That's been an incredibly exciting and profitable endeavor for us.

In other situations, for example with Java, we haven't found any reason to incorporate the technology into the products. When we look at what our products need to do, we say "hey, you know, we can write that program in C or we can write that program in Java, it's going to do the same thing." And we've just kind of arbitrarily made a choice to go with the more traditional technology.

XML

Something's going to happen. We don't know if it's XML or not.

Right now some of these new kinds of products are only in the early stages of being introduced. They tend to be somewhat PC-centric. They tend to use a much more traditional communications method to talk directly to a PC. Most of our stuff today is just standard HTML based.

We need some better protocols for allowing devices to talk to each other, but we haven't identified what those are yet.

Venture Investments

We're actually very heavily involved in communication with venture financing firms. We do a number of things. S3 does do quite a few venture investments, typically in strategic partners whom we believe will provide technology for products that we can produce in the future. We do that together with the venture capitalists.

Venture capitalists are often aware of many companies that we don't have contact with as well. Often times they will bring potential suppliers to us as a strategic opportunity so we can work together. In other cases we'll find a startup company that looks like it's going to be promising two or three years out. We certainly don't want to fund the complete development of that technology, so we will start to work with that company, sometimes as an investor, sometimes not. And we'll make introductions to some of the key venture capitalists and let them make the investment.

So it's a very back and forth kind of relationship where we each see ourselves as having a slightly different role in terms of stimulating development and new technologies, but our objectives are aligned. We want to get new companies started and we want to make them successful.

The VC Effect

They have changed the corporate strategy somewhat. Because of the abundance of venture capital, it's very easy to start a new company. So when we look for new opportunities, we tend to lean toward opportunities where our infrastructure's going to give us a critical advantage.

If five guys with $20 million can satisfy a market need, then that's probably going to happen.

So we would tend to focus a little more on the opportunity where the market needs a large tech support operation and it needs a broad sales channel, and it needs operational expertise. Those are all things that we can bring in very quickly, and that's something that a startup with venture capital can't duplicate.

So we would try to steer clear of some of the opportunities that venture capitalists can enable because we don't have a strategic advantage in those opportunities. We would focus more on the kinds of things where a more established company can excel.

I think anything that increases competitiveness influences corporate strategy in positive ways. They have shut down certain opportunities for tech companies, but in doing so have made us sharper, have made us more focused on the opportunities where we really can be very successful and contribute a lot.

The Future of Cable

We think it's going to move from one-way communication technology to two-way communication technology. We think it's going to be one piece of a very critical, broadband connectivity infrastructure. And we think that internationally it's going to catch up very, very quickly. We think that we'll see enormous growth the next couple of years, especially in Europe as there's big investment in infrastructure.

AOL Time Warner

Obviously they're going to push cable more toward a data-centric view, and I think that's good. We think there are more opportunities with digital technologies than there are with analog technology. There are more opportunities with broadband two-way communications than there are with one-way communications. I think AOL will help move people toward that model.

The Single Protocol

It will certainly accelerate the growth of products that work with that technology.

We have a very U.S.-centric view of this. The rest of the world already has a single protocol. And I think that because of that, interest in developing new interesting applications in the U.S. has been reduced, because people aren't focused on the global market.

So GSM has stimulated many more interesting new products and new services because it is almost a global technology. I think it's in everybody's best interest that we move toward the next generation global technology. The service providers, equipment providers, and consumers. And it's going to happen. There's no reason for us to have multiple technologies going forward.

I think the single protocol is going to expand the U.S. market greatly. The U.S. market is very fragmented now. One of the reasons people hesitate to invest in mobile technology is because they have to produce three or four different flavors. If somebody can produce a new device, a smart phone or a smart PDA or an automotive traffic monitoring system that works worldwide, they're going to be much more heavily incentivized to make that investment if it can work in multiple countries.

Look at who's going to be pushing to develop those technologies: the big communications companies like Nokia and Ericsson, the big automotive companies, big consumer electronics companies. Those are all global companies, so they're looking for global solutions.

Broadband and Satellite Access to the Internet

Most of the industrialized world is wired, and it's going to be wired with fiber. And then what we're going to have, in addition to wired networks, is local wireless connectivity through local area wireless networks, where you can move from one network to another. And that is technically a much more straightforward solution and much less costly. I think that's what's going to dominate.

The Future for Internet Telephone Service

I see it so much that I don't even think anybody's going to think about it. I think that, as I said, you're going to have ubiquitous two-way data utility in your home, and for a very low cost without thinking about it, you're going to move all kinds of data in and out of your home from dozens of different kinds of client devices, and one of those is going to be a telephone. At some point not too far out, it's going to seem very anachronistic that you run a separate analog wire to your house when you've got 10 or 20 megabytes per second of broadband, digital bandwidth.

The OS Battle

The OS battle for the PC has been won, right? I don't think there's any question that Microsoft is going to continue to be extremely successful in the PC space. The question is what's going to happen for things that are not PCs. What we've seen is that the OS market opens up every time there's a platform transition. Many generations of computers have had new successful OS's. PCs grew around a new OS. Appliance devices are going to have new OS's. The OS model is going to be very different. We have to see what's going to happen.

If you'd asked me six months ago I would have said that Microsoft at $35 a copy for Windows CE doesn't have a chance. But Microsoft has started to adapt to this market. They've started to realize that they can't push Windows models onto an appliance market, but they have to adapt their software and their business models to what that market demands. What that market demands is very high reliability and very open access to interfaces, at very lost cost in volume. Microsoft seems to be catching onto that and we'll see how quickly they adapt.

End-to-end Internet Service Companies

I think there are companies that can do that. The question is whether or not anybody should do that.

The big change in the networked world is that you realize that being here is no different from being there. The advantage to doing two things together is often very small compared to doing two things separately, because with networking the two separate things can talk to each other as well as two things that you've done together.

A lot of engineering tradition is to try to integrate things, whereas in a properly designed networked world things that are widely distributed work just as well. I think what's going to happen is you'll start to see best of class solutions working together. There's no incentive for somebody to provide more than one service or component of the infrastructure unless they can be the best at both of them, and that's unusual.

I think you're going to find lots of alliances, lots of services and companies working together, but not one company or one service starting to dominate across multiple areas. Because, for example, it's not clear that being the best at providing storage gives you any advantage in terms of being the best at providing bandwidth. So those may be separate providers.

The Effect of the Market Shakeout

It means going back to fundamentals. A year ago we saw a lot of business plans that said we're going to start this company and we're going to create a lot of excitement, we're going to IPO at an extremely high stock price and somebody will buy us, or we'll all get out of the stock and take our money and run and then go do something else. That's just not going to fly anymore.

The only kinds of businesses that are going to be successful are the ones that say "hey, we're going to develop a technology or a business model that in the end is going to allow us to make a profit." I think that people had the misconception over the last couple of years that perhaps that was not part of the whole Internet strategy. The Internet strategy was about creating stock price. But in fact many of the companies that have been involved in the Internet really all along have had a strategy of creating value, of creating commerce, of creating services that people will pay for. And those will be the survivors. And people will start looking at more traditional business timeframes—two, three, four years to create revenue and profitability—rather than these flash-in-the-pan three and six-month business plans.

Exciting Aspects of Being a CTO

The most exciting part is being connected to lots of people who have new ideas, and being able to take independent new technologies and put them together so that they're more than the sum of their parts. For example, for us, the most exciting one recently has been taking the home networking technology and the music technology, and being able to show people how to use them together. You can do incredibly cool stuff with music in your home that you couldn't do with just one of the two technologies separately. I think that's the most exciting thing: being able to go out and take ideas and put them together and find synergies.

The Dream Technology

What I would want is a computer in my house that can anticipate my needs. I want to be able to have a computer that knows what's going on in my house, that knows what's going on out in the rest of the world, that knows what kind of information I need, what kind of things I need, what kinds of services I need in the house, and can make them happen for me. Unfortunately, we don't know how to do that yet.

Andy Wolfe was appointed Chief Technical Officer in May 1999, prior to which he served as Director of Technology. During his tenure, Andy has successfully integrated various technology divisions associated with the acquisition of Diamond Multimedia, managed the development of the Savage/MX and IX mobile 3D graphics accelerators, and directed several strategic investments and acquisitions. Before joining S3, he was an Assistant Professor at Princeton University where he taught Electrical Engineering for six years. He is currently a Consulting

Professor at Stanford University and has been the recipient of numerous industry awards including the AT&T/Lucent Foundation Research Award. He has received four patents, has been published in over twelve journals, and has presented over 40 conference papers. Andy holds a B.S.E.E. from Johns Hopkins University and M.S. and Ph.D. degrees in Computer Engineering from Carnegie Mellon University.

NEIL WEBBER

Building What the Market Needs
Vignette
Former CTO, Co-Founder

Background

I've been in the software industry since I graduated in 1983, and I have worked in a succession of startup companies starting with Epoch Systems in Massachusetts, a hierarchical storage management company that was eventually acquired by EMC. I also worked for IBM here in Austin. The other startup companies I worked for included Dazel Corporation here in Austin, and Vignette of course.

Area of Expertise

My real background is in systems management and systems software design; big, complex systems that have to integrate with lots of different existing legacy systems and lots of other subsystems. Epoch Systems was basically a storage management software, systems management company. Dazel was actually founded by the guy who was co-founder of Tivoli. Dazel was another systems management kind of product. I worked at IBM as an architect in the AIX storage management area. So my area of expertise is big, complex systems that need to manage lots of information, and interact with lots of subsystems.

Being at an Established Company Versus Being at a Startup

The biggest difference is the rate of change and the flexibility required. Successful startup companies are usually positioned, by definition, in early-stage markets where things change rapidly. So the successful startup companies change rapidly.

We used to joke a lot at Vignette that the good old days were two months ago. In a larger company people might talk about how things were different three years ago, but at a small company you might go from 50 people to 200 people in a span of six months. That's a radical qualitative change. For example, in February of '96 Vignette had two employees. Four and a half years later it has on the order of I think around 2,000 employees. It has offices worldwide. It has multiple development offices across the country. And for all of that to happen in just four years—if you think about it, 12 months ago was 25 percent of the life of the company. So I think the pace of change and the flexibility required is the key unifying difference.

Being a CTO

I'm actually no longer working for Vignette now on a day-to-day basis. I resigned my role as CTO of Vignette in October of '99. I have basically left the company to pursue other interests. I'm pretty active now in the Austin area advising small-stage startup companies. I work with Austin Ventures Labs, called AV Labs, as an advisor to seed-stage and younger companies.

But to answer your question about how my role evolved over time at Vignette: as a co-founder, obviously, I started out with responsibility for the vision of the company and getting the company going, and my specific responsibility was running the development team. I hired the initial team and ran it until it got up to about 20 or so people.

As many technical co-founders do, in the early days I actually even wrote some code but that didn't last very long because pretty soon the demands of being a co-founder and being at the executive level just made it impossible to keep up with that.

Probably about a year into the company, as it started to grow and really take off, we went and hired what I always thought of as a professional management team. Ross and I had a certain amount of experience and we certainly knew how to build groups and manage them, but we had no experience in running an entire company. So again, like I think most successful startups need to do, we went out and sought professional managers who knew what they were doing so that we could focus on what we were good at.

We hired a VP of Development. We hired a whole executive team, and I took over the CTO role so that I could concentrate on our overall technical strategy and direction. You do hear this story a lot, but I think it's important to add to this that one of the things I believe we did very well at Vignette was knowing what our limitations were. We were not out to run the company because it was our plaything. We were trying to build a successful company and wanted to do whatever it took to make that happen.

The Role of the CTO

I think the most important role of the CTO is to be what I think of as the translator between the executive level management of the company and the technology aspects of the company. That seems to be bi-directional.

Here's a silly example I often used, kind of an absurd example to illustrate the point. If the marketing department says gee, if only we could make a Star Trek transporter, imagine how much money we could make. Here's all the people who would want it, here's how much they can pay for it, yada, yada, yada. Obviously you can't build a Star Trek transporter, but it's possible to get the entire company very excited about the possibilities of gee, imagine what we could do with this thing. Your role as CTO is to say well, we can't build a Star Trek transporter, but here's the thing we can build. You should understand how that would meet the requirements of what the company was trying to do.

I think of the role as two-fold. It's being able to translate in both directions from the executive to the technical guy. It's also a matter of translating the technical guys who may be coming up with some neat stuff but don't understand how to position it or how it fits into the company strategy.

That's sort of the day-to-day role of the CTO. Then I think a CTO should also focus on the high-level strategic technology developments and how your company needs to be positioned for that.

What I find is that a lot of CTOs don't operate on that high level; they're much more comfortable playing the role of what I would think of as chief architect. They're down there with the technical guys helping them design a database scheme, or breaking the systems up into modules, or doing what would basically be senior architect level type things. And there's certainly a need for that in a company, but I don't personally view that as the role of the CTO in a software product company. You need to be operating on a much higher level and let the guys who are in the system on a day to day basis do that other kind of work.

How the Role of a CTO Changes When it's at a Content Oriented Web Site

In that case the CTO is more likely closer to a CIO kind of role, they're the one who's designing the overall selection criteria for the products you're going to buy and the architecture you're going to fit within, that sort of thing. That's obviously a different role.

I have no direct experience with that, but I think of it more as a CIO kind of role in that case. You're selecting products, you need to understand the market and what's out there, you need to recognize whether you need to be a translator at all, and you need to be a domain expert in what it is that your company needs. You also need to be able to translate that out to the vendors you're dealing with and the people that you're looking for, the integrators and such that you're working with and the people who are providing you with the solutions.

Interaction Between CTO and CEO

I think it's important for successful companies to have a balance between their technology aspects and their non-technology or marketing aspects. And I'm going to say something that you'll probably find surprising. I actually believe most companies focus way too much on the technology and are too enamored of the technology-driven aspects of the strategy. What they ought to be doing is looking for problems the market has and trying to find the easy ways to solve those problems. A lot of times companies look at it from the point of view of hey, here's this really hard thing that we built. Where can we go sell it? That's exactly backwards.

But the general answer to your question is I think the interaction between CTO and CEO needs to be that balance of what can we build, where is there a need in the market, how can we make some money. If you want to build a successful company, how does all of that integrate into a coherent strategy?

Responsibilities as CTO During Fundraising Events

I think as far as fundraising goes, my role was probably much more strongly colored by being co-founder rather than CTO. But the generic answer to the question goes back to the translation aspect again. It was really my job as CTO to help the venture capitalists and whoever else understand our market, what our products do. I had to be able to explain things to them in ways that made sense to them.

We often would get asked questions that would come from people having too shallow of an understanding of what's really going on out in the market. And I don't mean shallow in a derogatory sense, I just mean that in a descriptive sense – venture guys don't have the time to get a deep understanding of every specialized technology area.

In fact, go to any four software company Web sites today and read their marketing brochures and they'll all say "our product helps you make more money over the Web." They all boil it down to that level of abstraction. Sometimes they'll say things like "our products will help you attract more customers, help retain customers." There are different buzz words that everybody goes through and they all copy from each other and it makes it look like everyone's product does the same thing. Because of that level of abstraction, you may get questions during fundraising like well, don't you guys compete with Company X? And as a technologist you might look at that question and in the back of your mind think that's got to be the stupidest question I've ever heard. The products are totally different, they do totally different things, we have 19 examples of customers using both products. It just makes no sense to you. But you need to understand the venture capitalists' level of understanding, and it's not limited to them. Basically anybody who's not involved in the industry in depth will have a much lower level of understanding, and you need to find ways of explaining what you do to help them understand all these different components without overwhelming them with technical details.

Building a Technology Team

I would say that goes in phases in a company's growth. In the early days we actually focused a lot more on just hiring intelligent, competent, experienced engineers. We didn't try to look for people with specific experiences in specific areas because, like I mentioned earlier when I was talking about what's different at a startup company, everything changes all the time.

If I went and looked for a developer that had a specific expertise, it could very well be that three months later I didn't need that expertise anymore, and hopefully if I hired him there would be something else that he would be good at. So we always looked at just general maturity and ability to deal with a startup environment.

In the early days, we frankly tended to hire much more experienced guys than a lot of startups do. A lot of times people look for the kids right out of school who will work 24 hours a day. Personally I preferred to hire people who had been in the industry for a while and who had the experience and who could deal with all of the changes that we were going to go through as a startup.

Deciding Which Resources to Devote to Emerging Technologies

We always looked at these things from a fairly ruthless point of view, asking did they or did they not help us get closer to our goal in terms of the product we were trying to build. Take Java for example. Vignette was actually a very early developer with Java, and when we looked at Java what we liked about it was the platform independence, the ability to deploy technology across MAC and PC and Unix.

What drove our decision to go with Java was entirely those kinds of benefits, and truthfully had nothing to do with the fact that the technical guys like it because it's cool or it's fun or it's a new thing or it's exciting. If Dixie cups and string would have helped us further our goal, that's what we would have used.

So we always look at the new technologies from the point of view of whether they help further the business objectives of what we're trying to accomplish.

General Trends

I think what you're seeing in the industry right now, in Vignette anyway, is a gradual transition to the next phase of the market. It's transitioning out of the early adopter, early-stage marketing, into a more traditional market. You're seeing a lot of consolidation. It used to be that there would be 19 different technology companies for 19 different bits of functionality that you would need in your Web site. What you're seeing now with Vignette's acquisitions and then other company's acquisitions is these things are being consolidated down into a fairly standard platform, tools, applications kind of play.

So I think in the Web infrastructure space, the days of hey, we've got this cool idea, here's this neat little feature, let's go build a company out of it—those days are probably gone.

You're starting to see that early phase gear up in some other markets, like wireless. The game just replays itself over and over again. Every time you get a new market you get 19 different companies springing up in that market to serve specific points and issues and they eventually all get consolidated.

Wireless Technologies

I think that wireless' technology is less interesting than what it means to the market overall.

Let's talk backwards for a second about the Web. There's really not a whole lot technologically interesting about the Web. The http protocol, the html format, all these things, they're built on ideas that have been there for a while. The genius about it, to give Tim Berners-Lee and those folks their credit, was to make it simple enough to the point where it could be widely deployed and everybody could use it. It was the ubiquity of it that really was where the power came from.

Looking at wireless, I would answer your questions the same way. It's less about what's cool about the technology. It's really more about how you can use ubiquitous connectivity: if you're constantly connected, if all of your communication devices can get you context-sensitive information no matter where you go, how is it going to affect society? How is that going to create new demands for products? What are the opportunities in there?

Keeping Vignette Technologically Nimble

I think the important thing is the culture you establish as a company. Ross and I founded Vignette based on the thesis that basically we don't know what we're dong, so let's go call everybody we can get our hands on and find out what they need and build solutions that address real customer needs. That's how we founded the company, that's how we ran the company. And that inherently just builds this nimbleness. The company exists and is structured to meet the needs of the market, so as the market changes the company changes. You don't lay down a five year road map and say here are the things we're going to do and here's when we're going to do them and view changes as something to be avoided. Instead you have no road map and you say let's go figure out what the market wants and then go do that.

The Recent Shakeout of Technology Based Companies

One of the things I tell people is that if you look at the NASDAQ, and look at the stock market, and look at the performance of various things, you really have to consider the past six to nine months or so as an aberration, not the norm. Things were going through the roof, and practically any silly little idea that somebody came up with-like myfavoritesocks.com-could get venture funded and possibly even go public in less than a year. That's the aberration.

If you return to what the historic norms are, the key thing is that good companies have solid, long-term plays, they have solid reasons for why they exist and why they need to exist. You need to get people to focus on that point of view when they are starting companies or selecting a company to work at. You need to focus on those companies that have a long-term success horizon.

Excitement of Being Co-founder and CTO

The most rewarding part was being able to build something, meaning a company in this case, that had lasting and tangible value. One of the frustrations of a software career is that a lot of what we build is very ethereal. It's hard to touch it, it's hard to see it. A lot of the programs that people develop, five years later there's no trace of them anywhere on the planet. It's all very throw-away, in a sense.

What was most rewarding about the company experience for me was actually building something tangible. We built something that is changing people's lives, creating wealth for the people who work there, helping companies go online, helping to shape the face of what the Internet looks like in terms of building successful businesses. That's just personally very rewarding.

Differentiating Vignette in the Marketplace

What's important about Vignette, was that we were looking for technologies in the marketplace that met customer requirements, met customer needs, and we would integrate those into our platform system. We were inherently always looking for the good ways to build integration points, or ways to build flexible systems so we could keep up with the ever-changing demands and the market and the customers. But I'd be hard-pressed to point to a specific technology and say oh, well we succeeded because we use Java or because we use [C++] or because we did this database layer. It's really not at that level. It's more of a philosophy.

The Dream Technology

I used to joke that the technology I wanted the most was kind of like whiteboard editing software; that after you've drawn this big thing on the whiteboard you could start to drag and drop the pieces around and manipulate them. But I actually think somebody's making that now.

Advice for an Internet Startup

You can probably predict how I'm going to answer this because I've said it 17 times already. I would say that the market tells you what you need to build. I would say, and I've said this to people at Vignette, that the absolute best idea is something that is really easy to implement, but everybody wants it.

So the answer to your question is to not focus too much on the technology. Focus on what the market wants and how you are going to build a successful company.

Future Technologies

I think the generic answer to that is you always need to ask what problem a given technology is actually solving. A lot of times you find things that people get very excited about, the press gets all wound up about them, but after six months or so the hype sort of dies down and the question is, where is technology X? The question is, what problem was it actually solving?

Maybe a good example from the past would be Push. If you remember Push and Point Cast and all that stuff, we were very excited about it. You've got to ask yourself the question, what real problem was that solving? And there are some real problems in there. You still see today there are elements of those technologies that are in use, and there are still companies that are doing very well with those kinds of technologies.

But they succeeded when they found an actual problem that needed to be solved, not what's cool or what's kind of gee whiz.

So I would answer that question by returning to that principle, for any given technology that you look at.

DWIGHT GIBBS

Let the Business Dictate the Technology
The Motley Fool
Chief Techie Geek

Background

It's a long and sordid tale so I'll just give you the highlights. I went to the University of Virginia as an undergraduate and picked up a Bachelors in Finance and MIS. I stayed and got a Master's in MIS. So I have some book learning. At the same time I was touring in a rock band. We were going to be huge so I looked for a job that would allow me to continue to tour with my band. I found one writing software for a small consulting firm in Charlottesville, Virginia. There I wrote software and did some project management for almost five years. In short order the band fell apart, I married my girlfriend, and I enrolled in Georgetown's MBA program.

During my first semester at Georgetown, one of my buddies at AOL tried to convince me to do an internship with a cool, new company called the Motley Fool. Given that the work sounded interesting, the Fool office was very close to my home, and I was a poor, starving student, I said, "Sure – sounds like fun."

I think I was the fourth employee. My job was to handle all of the technical back end work. Back then we were tiny – five guys in a shed behind one of the founders' house. We grew a little and used to joke about becoming a "real" company. To our astonishment, we soon became a real company with several employees and lots of customers. Between my first and second years in business school, the founders offered me an equity position in the company. As I did not want to work for The Man in Corporate America, I figured, "What the heck - I'll go for it."

You have to remember this is back in 1994-1995. There wasn't the Internet hysteria back then that you see now. So it was a pretty risky move. My parents trusted my judgment but I am sure they took some abuse from their friends – "I hear your son is a Fool <snicker>." I imagine they had some thoughts along the lines of "What are you thinking??!! Joining this company with such a silly name on this Internet thing??!! Can't you are least talk with the nice people at McKinsey?" But I decided to stick with it after graduation because my wife had a good job and if anything bad did happen, there were plenty of other things I could do.

Basically, I started by handling all of the technical aspects of the business – from desktop support, to database administration, to networking, to programming, to light custodial work, to everything else. I used to joke that I was responsible for anything and everything that had electricity running through it. Now I have some help. I am in charge of the Fool's technical staff of 110. We're known as the TechDome and are responsible for the Fool's web sites and internal technical operations in the US, UK, and Germany.

How the Technology Team has Grown

As you might guess, we grew very organically. In the old days, once my stress level got to a certain point, I would hire another techie to help me out. I assigned our web site operations to the first guy I hired so I could concentrate on our AOL site. At that point, almost all of our business was on AOL. That's where the revenue was so that was where I focused my attention. Eventually, I couldn't keep up with the internal tech operations (user support, networking, PBX support, etc.) so we hired another techie to take that over. Essentially, we hired techies to address wherever we were bleeding the most. Finally, after we got to about fifteen people, I realized that this growth strategy was somewhat less than optimal. We needed some organization and a growth plan. With that in mind, we separated the TechDome into functional groups, appointed team leads, and put together a rational growth plan. As we've grown we've had to reorganize a couple of times within the TechDome.

The Motley Fool has a fairly unique, non-traditional work environment. That is reflected in the titles and names of our tech teams. My title is Chief Techie Geek. The seven tech groups are:

The DORC (Developmentally Oriented Research Crew): This is the tech R&D group and is modeled after Xerox' PARC.

The Inter-Galactic Lords of Doom (aka InterGLOD aka IGLOD): In most companies, this would be called MIS. They handle desktop support, server support, network engineering, security, etc.

The DEAD: This is my Database Engineering Analysis and Design group. They handle all things database.

WebDev: This is my web development group.

AppDev: This is my application development group.

The Fifth Elephant: The 5E handles quality assurance and technical documentation.

The TechTrain: This is my technical training group. They offer classes for Outlook, Excel, Work, PowerPoint, Access, HTML, Project, several internal apps and anything else for which we need training.

As you can see, I gave my folks a lot of latitude in naming their groups.

The Role of CTO at The Motley Fool

My role encompasses many areas. I'm basically where the technology buck stops. One area of focus is technical strategy. I'm the one who translates company goals to technology goals and guides the determination of the technical strategies to support those goals. Part of my job is to communicate the company goals to the TechDome and show how our tech goals support the company goals. I have seven managers with whom I work very closely.

They are responsible for translating tech goals to the group and then individual goals. But I am responsible for keeping the TechDome aligned with the rest of the company.

I also spend a significant amount of time working on vendor relations. This is because we don't really have vendors. We have partners. We're not like some large companies who send out RFQs (request for quote) for everything they buy and go with the lowest price. We try to find companies that we can really work with and get them to customize their solutions for us. Some examples are Akamai, Cisco, Compaq, and Microsoft. We've worked with all these companies very closely and then they've been very helpful in solving our problems. I spend a lot of time growing those relationships.

The rest of my time is spent on project management and that elusive thing called "management." The details of management are really hard to describe but they are critical. It deals a lot with employee development and individual alignment. Day to day it means you send and receive lots of email, talk on the phone, and go to lots of meetings.

Areas of Expertise

When I started, I was doing all of the hands on tech work – database administration, programming, networking, hardware/software support. Happily for the last two years I haven't written one line of code. In fact, when the last bit of my code was removed from production, there was much rejoicing all around. While I haven't been doing the hands on work, I've maintained my knowledge of what's going on, what we're doing, and what's possible. I have become pretty adept at determining how technology can support the business and making sure that we have the technical resources we need to address the business needs that we have now and those that we will have over the next 12 to 18 months.

I guess one of my areas of expertise is that "management" thing. Through management, I make sure that the TechDome addresses the needs of both our internal customers (the people within Fool HQ) and our external customers (those on our web site). I try to get my folks to "manage expectations." If you manage people's expectations, no one gets upset. They know what to expect and are not surprised. That applies to both internal and external customers. Internally, we need to deliver projects when we promise to and make sure that they are to spec.

Externally, if we are doing maintenance on our site, we need to tell our customers well in advance that certain parts of the site will be inaccessible. I have to do this within the TechDome – "These are the projects we will be expected to deliver."; "These are the tools and languages you can expect to use."; "I expect you to put in X hours a week."; "This is the kind of compensation you can expect." It all comes down to making sure that everyone is on the same page. It sounds incredibly simple, but it's not.

Deciding What the Best Technologies or Vendors are for The Motley Fool

We are very focused on analytical, fact-based decision making. There are two words that I hate to hear when we are evaluating products/technologies: think and feel. When I hear, "I think …" or "I feel that …" I am immediately skeptical. Decisions should be made based on facts. I would much rather hear, "Given that ISP X is 20% faster than ISP Y, we should go with ISP X."

To get the information we need to make decisions, we conduct bake offs. In a bake off, we ask each vendor to bring their technology into Fool HQ and show us how it will work in our environment. This allows us to accurately gauge which products are best for us. We just rolled out Tivoli and BMC for our enterprise systems management. To determine which vendor to select, we asked Tivoli, BMC, and Computer Associates to spend a couple of weeks in Fool HQ showing us their products. They came in and installed their software on a test LAN. After giving us extensive demos, they let us use the software for a couple of weeks. Based on this hands on testing, we determined that Tivoli had the best overall framework for our environment and BMC had some superior point products. So we ended up with a hybrid approach and are using both of them. Another example of this is our ISP evaluations. During these tests, we install servers in different data centers and run tests using Keynote and SiteScope.

In order to best gauge technology, I also participate in a lot of networking. I talk to a lot of people in the industry - reporters as well as other CTOs. In fact, I recently returned from a CTO's conference in Colorado. You learn a lot just by talking to people doing the same things you are. You see what problems other techies are running into and how they are fixing those problems. You can also get a good read on who other techies are talking to, what vendors they like, and that kind of thing. I very rarely send my techies to training courses. I much prefer to send them to conferences where they can talk to people who are doing the same thing they are. That's where you learn the truly valuable information.

How the CTO Interacts With the Management Team

We have a CEO who is ultimately responsible for the business, but we have something called an executive committee that basically provides information for the decisions, sets the priorities, ensures execution and that kind of thing. I sit on the executive committee, as do our two founders, Tom and David Gardner, our CEO, our CFO, our chief legal officer, our COO, and several other top-level managers. We determine what's strategically important to the Fool and set the company goals. I inform the decisions from a technology standpoint. I am also the tech liaison to the other departments in the Fool. Once the executive committee sets the company goals, it's my responsibility to take them down to the tech level and determine what we need to do to support those goals. It's that alignment thing.

Keeping a Large Staff Organized While Fostering Creativity

It's pretty simple - we have to hire excellent technical managers and delegate responsibility to them. Keep in mind that technical managers aren't the same as regular managers. Obviously, technical managers have to be good managers. They also have to understand the technology. If they don't, their staff will not respect them. Techies can sense incompetent poseurs and good techies will not tolerate them. So if you're trying to pose, you're just going to get slammed. And it is likely to be exceptionally brutal. For this reason, when I'm in meetings with some of my staff and we're discussing different technical solutions, I'll often say, "Clearly I'm not the best person to make a decision on what the best technology is. That's why you're here." And that is the truth. I've hired very smart people, so it's up to them to say what they think is the best technology or best approach. I'm more concerned with the business end. I'm the one asking, "Does this make good business sense for the Fool?" For this reason, we don't always do exactly what we want to. It has to be balanced with the company's needs. If we did everything we wanted to do just from a technology perspective, we'd take all the money the Fool has and then some, and spend it all on tech. But you can't do that and be a sustainable company. It's my job to make sure that we balance the business needs with the technology needs.

The best way to foster creativity is to give people as much latitude as possible and encourage bold thinking. We also have a pretty liberating environment that encourages collaborative thinking. We don't have offices. We don't have rigid policies: no set business hours, no set vacation time, no set sick leave, no dress code. We also have a well-stocked game room. In short, we treat our employees like adults and expect them to act like adults. They generally do. This relaxed environment attracts creative techies and stimulates their creative juices.

Becoming a Leader

I think there are two basic traits that help people develop into leaders at any position within a company. Ironically, they are two of The Motley Fool's core values. The first is *innovation*. Good leaders will try new ideas and search relentlessly for better solutions. The second is *honesty*. Leaders have to paint a true and realistic picture of the way things are and bring uncompromising honesty to all of the things they touch, great and small.

Important Technologies

One of things I'm really excited about is wireless. Wireless is huge in Japan and Europe. And it's getting larger here in the U.S. The Motley Fool sees wireless as a significant distribution channel for us because it is a two-way medium. It's interactive, but, unlike the web, you don't have to be strapped to a terminal in your office or in your home to actually use the wireless technology. Wireless will allow our customers to take the Fool anywhere and everywhere. It has what we like about the newspaper - you can fold it up and take it with you. In addition, wireless gives us the two-way interaction that the newspaper never will. It's the best of both worlds.

Obviously the biggest problem right now is the interface. Brutal honesty is that most of these things are completely unusable for two-way interaction. But that's going to change because there's a business case for changing. One of the things I'm really excited about is voice recognition on wireless devices. Right now I can theoretically use my cell phone to send a text message. However, that is a huge pain in the butt. It takes forever just to write a one sentence email. It's so bad, I don't even bother using it. However, if I could talk into my phone and have it do some voice recognition and convert my message to an e-mail, that would be tremendously valuable – incredibly compelling.

Using the Fool Technology to Make Money

A lot of people go on vacation without their computers, but they still want to keep on top of what's going on with their money. Let's say you're a daily visitor to the Motley Fool message boards. Via your wireless device you can stay in touch because you don't need your computer.

Another area where wireless could be quite useful is email. When I go on vacation, my wife is generally pretty adamant that I go laptop-less. But that makes it quite painful when I return. I was recently out of town for about ten days and returned to find about 2500 unread email messages in my box. Had I been able to go through my e-mail and respond to the critical issues while on vacation, it would have been so much easier for me when I returned. A wireless browser would have allowed me to stay current.

Technology That Gives you an Edge

Technology itself gives us zero competitive advantage. The fact that we use ASP or Linux or whatever means nothing. It's how we use the technology that really matters. The implementation is where the value is. What's our competitive advantage? It's primarily our brand. People trust us - they believe in us. We're not conflicted like the investment banks and the full-price brokers are. We also listen to our customers. Our customers tell us what they want and which solutions will provide them with the most value. Obviously you'll never find a large homogeneous group. Not everyone agrees with what you do. But if we give our customers the tools and the flexibility they ask for, we can really help them take control of their financial lives. That's what's key for us.

Resources in Demand

It's people. Technology in and of itself is worthless. It's how you use it. The funny thing is that right now the industry is not hurting for the traditional tech skills – programming and sys admin. It's dying for project management skills and really good quality assurance and testing skills. Project management is just a tremendous need because techies who can code are relatively easy to find. Finding someone who can manage a large project is a different story.

Our software has advanced to the point where it's not a case of one guy or one gal handling a single project. We have multiple techies on each project, and so someone who actually understands technology and project management and can execute and motivate is tremendously valuable and in short supply.

Another tough skill set to find is quality assurance testers. It has been in high demand because most techies approach applications thinking like a techie and not like a user. What we value in our QA folks is their ability to think like average Joe user, but find problems and be able to diagnose why they are happening.

Technical Skills That are in Demand

Data base administration. You can find lots of people who say they're a good DBA, but finding a really good DBA is difficult.

Monitoring the Industry

It sounds trite, but to stay on top of the tech industry, I talk to a lot of people in the industry. What I find really valuable is networking. Talking to other people in the industry about what they are thinking and hearing is tremendously important. There are also a couple of reporters who help me stay abreast of recent tech developments. I ask them what they're covering and what they think is interesting. Of course, you have to be careful when you talk to vendors because they have an agenda. Also, we subscribe to 28 different technology oriented magazines at the Fool and I read them all almost cover to cover. It is a good way to keep a pulse on what the general feeling is in the industry.

Important Industry Events

I've spent some time thinking about how the merger of AOL and Time Warner will affect our business: What is this going to do for us? For me, it's more about our business than the technology we use. We partner pretty closely with AOL, so it has to come back to the business. I know I sound like a broken record with my focus on the business before technology. That's because we are actually focused on building a sustainable company. This is pretty rare in the Internet space. You find a lot of people who are looking to cash out with the IPO. For us it's more important to build a lasting company. If we have our own cable TV show, perhaps we need to put together a state-of-the-art TV studio in Fool HQ. If we are going to stream that show over the Internet, we will need more bandwidth. Whatever direction our business takes, it's the business that drives tech, not the other way around.

How Things Change When a Company Gets More Money or Goes Public

Well, our budgets get a lot more scrutiny, I can tell you that. We're still private but we have taken one round of VC. Happily they own a small chunk of the company so we are still in control. With funding comes scrutiny and that is actually a good thing. When you have someone externally looking at you, who can easily compare you to other companies in their portfolio, it makes you more cognizant of things like a cost/benefit analysis for each project you take on. You start asking more questions like, "Are we really getting our bang for the buck?" and "Are we really providing a good service or would we be better off outsourcing some of the things we're doing?" Having external investors makes us much more cognizant of the ROI on all the tech projects that we take on. We have VC's who we really like and who have been really helpful. But, they want a return on their investment and we feel responsible to them in everything we do.

Keeping an Edge

I answer to our CEO – Pat Garner. He's not a tech guy. He's a brilliant businessman. In terms of business operations and marketing, the guy rocks. But he's not a tech guy and he's not supposed to be. For me, it comes back to networking again. I talk to people in the industry. That's why I go to CTO's conferences. I have been able to talk to the chief scientist from eBay, the CTO of DoubleClick, the CTO of CBS Sportsline and several other people like that who are running large sites. I find out how they are handling the problems that we are facing and what new technology they are evaluating. These conversations help inform our tech decisions. I also have a little cadre of geek buddies that I talk to. It's somewhat competitive. If I see one of my friend's sites down, I'll send him a little nasty-gram mentioning it. I would expect to get the same thing in return. A little friendly rivalry is good.

Dealing With Explosive Growth

Scale definitely reveals problems. A site that can handle 10 people may not be able to handle 10,000 people. You need to build your site to scale from the beginning. Unfortunately, we did not. Back in 1995, we didn't know what we would be expected to support. Nobody knew the Web would be as big as it is. For that reason, we have had to go back and gut many of our systems. Last year we completely redesigned the network and server architecture of the site so we could scale to meet the demand. Right now we are doing the same thing with our code and our teams. Had we designed it right upfront, we would have saved a lot of pain and anguish on the back end.

Changes When you Build International Sites

The biggest thing for us is that we have no maintenance window. It may be midnight here but in the UK there are plenty of people who are at their PCs. There are even more in Germany. Once we go live in Asia, we won't have a maintenance window at all. We have to be very cognizant of what's going on in different time zones as well as what's going on in different markets. If the German market goes crazy, we're going to get more traffic from Germany. The same with the UK.

There are also localization issues. Not everyone cares about dollars. Not everyone cares about Eastern Standard Time. The Brits are unique in that they put u's after o's in random words, like favourite. And they misspell color – colour. You have to be very, very cognizant of what people expect. Their whole date format may look screwed up in the US, but that is what they are expecting and you better deliver. If we are going to be in foreign markets, we have to look like natives. If not, we might as well close the site down. We cannot code for the US market exclusively. In everything we do we have to be cognizant of other cultures and how they look at the world.

Advice for a new Internet Company Focused on Technology

From the beginning, plan to be huge and international. We didn't do that. When we first started we never knew we were going to be this big. We never knew we would be operating in multiple countries. It's been very painful because we've had to go in and gut our hardware, our networking, our software – everything - because we didn't build that in from the beginning. That was back in 1995. We just didn't know what to expect. It would have been fairly easy to do. You don't have to build out a huge, internationalized site, but you have to have the framework in place to allow for it.

Making Your Site Look Professional When you Have Limited Resources

It doesn't take a lot to look professional – just some good designers, a few HTML coders, and a couple of programmers. The big expense comes when you have to scale up and handle millions of page impressions a day. If I expected 100 visitors a day, then I could easily build a site that looks and acts exactly like eBay. But to scale that up to accommodate tens of millions of people a day – that is exceptionally tough. So it's not so much the look and feel that's difficult. It's scaling a site that's tough.

New Types of Technology Emerging

Real voice recognition is pretty compelling. In the future, systems will understand accents and understand people at the rate at which they speak. Having systems that can recognize your voice, understand what you're saying, and act on voice commands will be quite valuable.

Broadband, real wireless and wired broadband is pretty exciting, at least to a geek like me. Once I have a cheap 100 megabit connection to my Palm Pilot, I'll be thrilled. That will enable me to do so much more than I can right now.

The other thing I would love to see is translation software that actually works. With that, we could just run a news story through a converter and get a German translation that actually sounds right to a German speaking person. That's ungodly difficult today. Particularly converting a story from English to something like Mandarin Chinese.

Someone behind me is saying, "Yeah and people in hell want ice water," but you never know…

Dwight Gibbs earned degrees at the University of Virginia (B.S., Finance and Management and Information Systems; M.S., M.I.S.) before writing software for a small consulting firm and touring with a rock 'n' roll band. After four years, he gave up his dreams of stardom and enrolled in the M.B.A. program at Georgetown University. Around the same time, Dwight joined The Motley Fool as Chief Techie Geek. As the only Foolish techie, he handled all technical aspects of the business. Six years later, Dwight now manages the Motley Fool's TechDome that supports the Fool's America Online site, websites, 350 node FooLAN and a 110 person technical team.

PETER STERN

Building Your Technology Solution From the Ground Up

Datek

CTO

Background

I have always been interested in building things. In high school I spent all my minimum wage earnings from frying hamburgers on an Atari 800, with a floppy drive (most people used cassette tapes at the time) and that pretty much sealed my fate.

I went to Carnegie Mellon, earned a degree in Electrical Engineering (never actually took any software or programming classes at college), and then spent 6 years in the aerospace industry working on all sorts of really interesting projects, mostly involving lasers and helicopters. And then one day my friend Josh Levine, who was working on high performance trading terminals, convinced me to come to NYC to try to build something I had been telling him he should build: take trading terminals and port them to the Internet so more people could use them. The rest, as they say, is history. Datek Online, after 4 years, is now the 4th or 5th largest online brokerage firm, due mostly to its superior products.

Changes in Technology

Technology has had a huge effect on this industry. Now, pretty much everybody demands access to an incredibly large amount of data. Take technologies ten or fifteen years ago—if you dump everything into a database and then let people access the database, it doesn't really work. People have come up with a lot of different approaches, like 12,000 option tics per second available to anybody who might want them. Think about how many trades happen in a second on NASDAQ. There are a few thousand different transactions executed in a second, and then you have to post those results to everybody—that's a pretty amazing feat.

It is completely different than ten thousand searches on Yahoo in a second, because ten thousand searches on Yahoo in a second is pretty much ten thousand viewers searching for the word "sex," so that is a comparatively easy system to build. The contents of the web that relate to "sex" do not change on a second-by-second basis. On Yahoo, you're talking about copies of the exact same data, but in the financial market one has to access personalized, rapidly changing information, so that's been the challenge.

How Financial Companies Have Evolved Compared to Other Companies Using Technology

Well, there are two different ways to talk about this. One is no, it's not any harder, because people are making the same technical innovations in the financial world as they are in other industries. Yet over the entire industry it probably has been harder because the financial industry is not one to take chances with technology. Many technologists on Wall Street are risk-averse. They've been doing the same thing for tens of years and there are accepted ways of doing things in all areas of banking and finance, and accepted ways to use technology to solve those kinds of problems.

There has been a lot of innovation at the call center but in the back office there's been very little innovation, especially in the way people solve back office problems. On the front end there's a lot more competition, so you see a lot more innovation.

It depends on how you split the problem up. For example, some people are still using the same computers and the same ideas to clear trades and perform back office operations that they've been using for thirty years. There are few large-scale brokerage operations, and they don't change very quickly. In front-end systems there's the whole day trading industry that's grown the last five or ten years, so there is an enormous amount of innovation in representing information to people and letting people choose different financial strategies based on that information.

Areas of Expertise

Well, I came from a military background, where I was building real embedded systems—basically things that would go in helicopters and that kind of thing. In that world everything that occurs happens in real time. When I came to Financial World, you didn't need to do anything in real time—that was the understanding five years ago. When you're flying a helicopter, several seconds later means you've already hit the trees. You can deal with maybe several microseconds later, several hundreds of microseconds later, but certainly not several seconds later.

What I did was take the concept behind day trading, where you can hit a key and get a trade done that second, and expand it so a customer could do something similar over the Web. The Web used to be considerably slower than a dedicated terminal. There were people building day trading terminals in the late 80's and early 90's that would let you very easily access electronic trading paths, where, for example just by hitting a key you would have the quote in front of you; by hitting the shift key, messages would go from your computer to NASDAQ and back to you, and you would see your execution a second later. I ported that concept to a web interface.

Role as CTO

I'm responsible for all of the technology behind the retail brokerage of Datek that you see, including the back office systems.

Our company is pretty much founded on technology so it is a really interesting job. Everything we do is technical, we are a highly technical company, but I think the novel part of what I do is I try to build organizations that are going to build good products. When I compare people that I consider good CTOs to people that are not so good, good ones understand that there are certain cultural and organizational problems that you can solve that are going to make people have a tendency to build things better. For example, technical decisions should be made on technical merits not on political ones. You have to be careful of that, and if you are forced to make a political decision you should advertise that to everybody—okay, this is not a good technical decision but it's a good political decision. Otherwise, your technical organization quickly spirals into producing least common denominator products that nobody wants to build, and, more importantly, nobody wants to use.

It sounds silly, but there are a lot of really good organizational and procedural elements that enable good people to build good products. That is the novel part of the job, and the hard part.

How the Technology Team Interacts With the Rest of the Company

Well, we are involved in every aspect of the company, and it is frustrating for the company sometimes because there are only so many technical resources we can throw at any given project. Sometimes the company feels like the technology team is a limiting factor and they should go elsewhere to get something done or build a product that doesn't involve technology, but of course that doesn't really exist anymore in our business. My technical teams are involved in all aspects of the company and a lot of areas of the company cannot proceed with evolving their business or making it more efficient at attracting new customers or building new products without resources from the techno-crews.

Keeping on top of Rapidly Changing Technologies

I don't keep on top of them all. Even back when all I did was one particular thing in the military I could not keep on top in just that one itty bitty little field, so at this point I don't even try. I cannot tell you who is making the fastest PCs, or who is making the fastest mainframes, or who is making the novel changes in database technology. The thing that I've noticed, though, is that it doesn't really matter that much.

First of all I have good people who are constantly pulling for the best products to fill their needs and I just make sure that they all know about each other. A lot of companies outsource a lot of things, for example a lot of their core products are built upon products that they've bought. A lot of our competitors buy pieces and then they put the pieces together. We have a simpler business, a more technologically focused business. We have to build more from scratch. We have a very clean architecture, its simple and elegant so it's easy for us to build highly scalable transaction systems. Our penalty for not being aware of the latest advances in middleware improvements or in high capacity hard drives is not that great. We push the envelope in architectural concepts, not with the actual platforms or technology.

The Advantage of Having Ownership Over the Whole System

Certainly a lot of our competitors are fighting their legacy systems, they need these pieces like shims to wedge in between other pieces that aren't really fitting together very well anymore. We don't have that problem. We have ownership over the whole system. We do not have any legacy components. We have that advantage because we built the whole thing from scratch over the last five years. Ten years from now we might be looking for a shim.

New Wave of Technologies Emerging

It depends on how you count waves. I don't think we are done with this wave. Computer technology has a ways to go before it is all pervasive and ubiquitous. Wireless access is going to improve enormously over the next few years. The products and services that I see evolving the fastest are all about making it easier and less expensive to access more information from anywhere.

Getting broadband into everyone's home and office is another small wave we have yet to complete. Once everyone has DSL you can build much more interactive products, much neater things, you don't need to be stuck with text. Right now we spit out bits of text to represent quotes, but imagine a world where you see three-dimensional objects spinning around in your screen. We don't do that now because the amount of data and the amount of processing time required is too high, but two or four years from now when everyone has a quad processor gigahertz machine and everyone has DSL…. Once this happens you will start seeing some amazing changes, radically different ways of representing information and doing things.

Is Technology Growing?

Oh yes, I think that technology in general has always been growing exponentially, you may not realize it as such but every day there is more technology released or developed than the day before. Computers already surround people in ways they do not appreciate. They have become like spiders: supposedly, you are never more than four feet from a spider. Once these computers can talk to each other, your car is going to start giving you driving instructions. You will not be booting it up or plugging it into the wall. You won't perceive it as a computer.

Incorporating New Technologies

I don't know if I have a two or three-year event horizon, although I probably should. I have a one-year plan that is rarely correct. We've always been very happy to have a six-month event horizon. We have a fairly small group. A lot of CTOs have thousands of people, computer centers that measure in acres, and they deal with computers that they don't even know how to fix anymore. We are not there yet and I hope to never be there, so we basically look at maybe six months to a year out because we're in the rapid growth phase.

We are not limited by technological limits. What limits us is people's time. Things that make people more efficient or more effective and reduce software development time, that's where we go, that's what we look for, things like Java. Yes Java is really inefficient when measured by CPU cycles, maybe it's even doomed for extinction, but it saves people a lot of time. A programmer working in Java can be much more effective just because of a higher level of abstraction than C. We do not measure efficiency in CPU cycles. We can just buy faster computers or more of them.

I don't know where we're going to be two years from now, but I do know that just when we are kind of worried that our networks are going to get a little overburdened and we are going to have to figure something out, Cisco or 3COM will come out with a network technology that is five times better than the last one. That's been a repeating pattern every year, which is well within the time frame that we need it to happen.

Programming Languages

Programming languages are funny things—there's almost a fashion to them, like clothing. The languages that we use are pretty much mainstream languages, I think Java's here to stay, C sharp is a neat idea, it's a very neat idea, in fact it builds upon C++. We will have to wait and see if it becomes the new fashion. We are trying to migrate into a world that is Java-centric, not because I think it's the best programming language technically, but mostly because it is widespread, so I can find people who know it, and because people can be more productive in it.

Resources in Demand

Companies like ours are limited mostly by their ability to enable business ideas, or resolve business problems, or make business solutions out of technology. There are not that many people who do the basic analyst's job of understanding the technology, understanding what the customer needs, and being able to tell the programmers how to build it, and that's still the limiting factor for a lot of companies, including ours.

There are a lot of smart people, great people, but the number of people who at a high level can determine good business vs. technology tradeoffs is still kind of limiting us. That's an experience problem, and the population of people who have those skills or experience at using those skills are needed by the industry a lot. I'm not talking about Web producers or programmers, it's always hard to find good people in anything, but those people who have that "je ne sais quoi," knowing how to build good stuff, without the clue factor, those people are pretty rare.

Skills Needed to Join a Technology Team

Anyone who has desire can learn any technical skill; it might take them a year, it might take them six months, if they've learned these kind of skills before it might take them a few weeks. I am not so concerned that they already have years experience with Java, I am not so concerned that they have years of experience designing Web sites, I'm concerned that when they look at a product or look at a service they see what it does and what it could do and understand how you can build the thing the way it should be.

The people that look at a cell phones and say, "you know what, if I designed this cell phone, I would rearrange the buttons, because I use this button a lot more than I use the other buttons so that button should be where my thumb is." Those kind of people should be able to apply the exact same talent to good Web page layouts, or determining which features on a product do not make any sense, or that the underlying architecture of a product does not match what it is supposed to do. They can do that whether they are talking about their cell phone or talking about Java code or they are talking about a Web site.

How Outside Funding Affects the CTO

It works out really well. When we are out trying to raise money, the technology portions go over really well, because our team has just done a great job building technology geared towards our businesses. We have a really good understanding of where our industry is going and what we need to build going forward. While I am not so concerned with the technologies used to build it, I am very much concerned with what we'll actually be building and we usually impress investors with our knowledge of this and our intensity, and with what we've accomplished. We have the second most visited web site on the Internet during working hours according to Nielsen's, and when it does have technical problems it doesn't have technical problems due to capacity.

What VCs are Looking For

Because of the press in '99 and '98 they're always asking about scalability and reliability. The number of Internet consumers is still relatively small—there are fewer than a hundred million retail brokers' customers trading online and they are not trading online with all their assets. Is our technology going to work if we have hundreds of millions of customers? That's what they're concerned about. They're thinking about several hundred million customers on a really big day in the market. It's not like the average online bookstore. A weird event may occur, and suddenly you have three or four times your average daily volume. We see this fairly often. Reliability and scalability, these are the two biggest concerns.

The People That Help Shape Your Vision for Technology

There are a few people that I meet with every few months and they usually have a lot of good ideas about where technology in general is going and what they're focused on and what their teams are working on. We exchange ideas. My original partner Josh Levine, who is the genius behind a lot of the technologies that we use and who now works at

Island is one. He is probably the single largest influence, he is a stereotypical guru, a wizard-behind-the-curtain kind of thing, and he usually has a good idea of where technology itself is going. The others are mostly founding CTOs or high level technologists at other dot com ventures.

I find it interesting that over the last year or two, more and more of our conversations focus not on specific technologies, but on how to build larger organizations where every individual is as effective as we were when we worked alone or in small groups. Three years ago I would not have expected the phrase "org chart" to come up so often in these conversations.

Handling a Crash or a Bug in the System

Our system is highly distributed. Architecturally, it is a giant tree, with a set of branches, and each system or computer hangs off this tree. Lets call each computer a "node." Every node works on a very small piece of a task, and needs to talk to other nodes to get something done. Every node knows multiple paths through the tree to talk to any other node it needs to talk to. A lot of nodes need to die before tasks can't get done. In other words, a lot of computers need to break at the same time before a customer will notice anything is wrong.

For example, if a Web server needs an options quote it has a set of computers to go to ask for a quote. If it cannot talk to one of them it just goes and asks another one. Those option quote computers are on different networks and in different places. Option quote computers are listening to different option pricing feeds coming from different data centers. If you go to a data center and started turning off computers at random, it's going to take you a while before you actually start knocking out any given service. All the different services are segmented so even if you do knock out any given service you're just going to knock out one service, like options quotes, but not hinder the ability to enter orders or get equity quotes.

This is the almost classical definition of a "distributed system." Distributed systems tend to be both reliable and scalable. "Scalable," simply put, means the quality of being able to add more capacity. In our system, to add capacity, as well as reliability, we add more nodes to the tree.

You have to keep the cost of managing all these relatively cheap, inexpensive computers low. If you can do this, then you can run a thousand computers and it operates more like a biological system. This is how the Internet works, ours is a very Internet-style system. When one of the routers on the Internet goes down you can still access Yahoo. When our computers die, you can still access enter orders.

These are really common ideas for people who do Internet systems and not very common ideas for people who do transaction systems. Transaction systems people want to look at a single spot on your hard drive platter and say, "that's the record, that's this guy's account balance, or that's the status of this order." Internet people are not concerned by the fact that I don't necessarily know in advance the path that my packets are going to take, and we do not care either. The real beauty to what we have done is that we have figured out ways of marrying the benefits of both those concepts. I do know the status of any given order but I also do not care how it got there and I do not care which hard drive the status lives on. That is the technical magic of what we do.

Industry Happenings Shaping the Future

The securities business is a heavily regulated business. Every aspect of what we do, whether its processing orders, delivering quotes, or even printing out monthly statements, is all controlled by government regulations. You can trade a lot of things on eBay, but you can't trade stocks there. Not legally, anyway.

Most of these regulations were created to protect investors in a time when information was difficult to access. Technology makes it easy to access information. Technology has changed the financial markets to such a great extent, that the original ideas behind most of the regulation we deal with are no longer relevant.

At the same time, changes in the financial markets have pushed technology to the limits, making it increasingly difficult for exchanges and brokerage firms to provide the basic serves upon which the entire industry depends. For example, decimal pricing of stocks has pushed the exchanges quote systems to the limit. Without quotes from the exchanges, it is difficult, if not impossible, to trade according to the regulations.

Technological change has made the financial markets more open, more competitive, and fairer for the individual investor. This change has caused the exchanges and regulators to radically rethink how the American financial markets should be structured. The pace of regulatory change over the last three years has been enormous.

Regulatory change is only going to continue to accelerate. The way that the markets work 5 years from now is going to be very different from how they do now.

It will be interesting to see how it all works out. Will I be able to compete on the basis of technological solutions? Or will new regulations make it irrelevant who has the most competitive solutions, by forcing a standardized way of doing things? Since my company works closely with the regulators on a variety of these issues, at least we get to voice our views in building the new world.

The Effect of the Government on Technology Growth

Regulations already impede innovation in the financial space. I'm a student of history, and I understand very well why the financial industry is regulated. There are a lot of good reasons for it. However, if we were creating a whole new framework today I don't think you would need to regulate it. There used to be enormous barriers in regards to sharing information, but the Internet completely changes that. It's now easy to share as much information as you want. You might have personal reasons or business reasons for not sharing it, but there is no technical reason why you cannot share all the information that has ever been collected.

Having experienced first hand how the powerful can manipulate regulation to squash competition, and having seen the innovative products we had to abandon due to regulation that was clearly meant to prevent other problems, and having to spend large amounts of resources explaining technical products to regulators, I see that regulation in this space is costing us all more than we can know.

I think every one recognizes this, and everyone knows change is coming. The regulatory agencies that we deal with on these issues are aware of the costs of regulation, and its effectiveness in an age where I can get a real-time stock quote on my cell phone. They are looking for some smart answers. It might take a little while, but changes are coming.

The Recent Shakeout in the Market

Oh, you mean the big bubble that burst? Well, it was healthy, it was needed, we were expecting it. I was always very concerned about the extreme speculation in Internet companies. So much for that new economy silliness. At some point you're supposed to make money right? It can't all be moonbeams and fairy dust. So while a part of me is crying, philosophically I think a return to reality is good. It certainly hurt a lot of companies that I work with, it certainly hurt a lot of my colleagues, and it certainly hurt us. It was needed, though; people need to be realistic about how they evaluate their financial investments, and in the long run smarter people and a smarter market will only help us all.

The Effect of the Shakeout on New Companies

The climate is terrible for raising money for small technology companies. A lot of value has disappeared. A lot of people are shell-shocked and investment shy.

But I am not sure that a lot of these companies have worthwhile strategies. Maybe they should not be started.

It won't substantially slow down the number of useful new technologies being started. If you have a good idea, people are going to respond to your idea, and you are just as likely to do it now, maybe even more likely if you don't have to compete with companies that should never be started in the first place.

Benefits From Separating the Technology End

Well, when Datek Securities and a set of consulting firms merged to become Datek Online Holdings, we were in so many different businesses that we just decided to split up what would normally be different departments into different companies. We build the technology for two different companies, and it didn't make sense to build two different technology teams and put them as departments in those companies.

It gives us a little more flexibility, too, because we can go and solicit outside customers if the work ever dries up with our main, internal clients. We can go, hey, we have a great team, we have a methodology, we have these skill sets, we know how to run these systems. At the moment it doesn't make much of a difference in our day-to-day lives. There is a separate cultural identity that is important for team building. While we certainly all work for Datek Online and we're all going to the Datek Online picnic, we also work in a technology company.

Advice for Internet Startups

I have learned a lot. I have lots of advice. Fortunately, nobody wants to listen to any of it. I'll see if I can be brief.

Be true to your vision, and avoid arrogance. Stay in the sweet spot between these dipoles.

If you are true to your vision, you will listen to customers and investors, but not too much. Never build what you don't understand, and don't take advice when it leads you down a path towards a least common denominator. If you try to keep everybody happy, you start with the best of intentions, and end up with a mediocre product.

Towards the other extreme, do not be so arrogant that you think you know more than you do. Think like Socrates, who claimed to know enough to know what he did not know. If you refuse to listen to anybody, you may end up designing products that only you want to use.

Oh, and don't be too clever.

And, of course, my favorite: use sunscreen.

Mistakes Technology Companies are Making

One problem I perceive is that technology development is still an art, and a lot of people do not want to recognize that and hire or deal with the artists, especially because technology talent is so hard to find and so competitive. A lot of people will neglect to hire that one expert technologist who really knows, who has that clue I was trying to explain before, and if you're not willing to go and get that clue person, you always end up with an organization that cannot really meet your technical needs.

I see that happening a lot: I see companies running into problems because the technical superstar, who might not have been a programmer but was the guy who could put it all together, leaves or gets pushed aside. Pretty soon the company fails because there is no technical strategy and vision. I guess vision is probably the acceptable word for this clue thing.

You need people with technical vision, and if you do not find people with technical vision then pretty soon you find that you cannot build new products, and you cannot maintain the products you have. I see so many companies struggling to get it right. Sometimes, I am not even sure they realize they are struggling.

The companies I admire most are the ones where there is that person or set of people at the top who clearly understand the business strategy and how the technical vision fits the business strategy. It's a pleasure to work with those companies.

The Dream Technology

You should be careful what you wish for. I'm a real pessimist, especially after reading that Bill Joy article, about how the human race is not needed for the future.

Learning is really tedious, yet I love to learn. And ignorance causes so many ills. Maybe it would be great if we could download knowledge, or at least content, directly into our brains.

Off the top of my head I think that this might solve a lot of the world's problems. But maybe the reason that learning is tedious is because you only get what you work towards, and it's the work that turns facts into knowledge. So perhaps this brain download trick would create a world of people who think they are wise but only because they know a lot of facts.

Ok, so maybe I just want digital paper. Digital paper could be kind of neat.

Peter Stern is a founder of Datek Online Holdings Corp. and currently is the Company's Chief Technology Officer. He is also president of BigThink, a subsidiary that develops and operates technical systems used by Datek Online.

He is recognized throughout the industry as a visionary who saw the commercial potential of the Internet in the early 1990s, when it was primarily used by engineers to exchange information. In 1995, he became a consultant to Datek, where he developed web interfaces for stock traders and automated trading systems. These systems were incorporated into the first version of Datek's advanced trading platform, which enabled investors to trade stocks on the Internet. This led to the June 1996 launch of Datek Online's brokerage service.

In developing Datek Online, Peter completely automated the broker's job of accepting and screening customer orders. He then built a real-time expert system to execute customer orders. This system minimized the use of traders and brokers and streamlines the process of routing customers' orders to the markets. Peter's goal was to use technology to give investors new capabilities for placing orders, not just to make traditional brokers more efficient.

Prior to moving to New York City and helping found Datek Online, Peter worked in the aerospace industry, where he worked on laser radar systems for helicopters. He has a bachelor of science degree in electrical engineering from Carnegie Mellon University.

WARWICK FORD

The Security Behind Technologies
VeriSign
CTO

Background

I've been in the communications technology field for over 20 years, starting in the Defense sector. Then I worked 9 years for Nortel, where I was responsible mainly for R&D and consulting related to cryptographic and security technologies. In 1996 I took a position with VeriSign as Director of Advanced Technology and that has since grown into the CTO role.

Major Responsibilities as CTO

There are at least four different responsibilities. One is to develop opportunities for exploiting new technologies in new products and services. Another is to help maintain and disseminate coherent technology strategy across our different product lines. A third is to work with technology partners or prospective technology partners on projects aimed at jointly exploiting new technologies. And a fourth is to drive outbound activities related to sharing our technology strategy and vision with customers, analysts and the public generally.

The Unique Technology of VeriSign

The core technologies are the cryptographic-based technologies of digital signatures and encryption, together with the transaction processing, database, directory, and distributed systems technologies that underlie the high end of today's Internet services. Our high-availability, high security service provider platform for issuing and managing digital certificates on a massive scale is unique and way ahead of the competition.

The underlying technology principles are not particularly new. However, there are continually new challenges in successfully exploiting these technologies within the application environment and the legal environment that surround their use. For example, the objective of many e-commerce participants is to be able to totally eliminate manual steps and paper-based steps from their business transactions and replace them with automated electronic transactions. This raises a number of questions, such as: Are the necessary records in place to be able to prove, in a court of law in the event of a dispute, that something happened or didn't happen? Maintaining such records demands some special technologies but also demands tightly secured and audited operations and sometimes certification or accreditation by regulatory bodies. Pulling together all these threads is really a challenge. Providing a scalable solution without customer organizations needing to make large capital expenditures up front is also challenging, and is key to opening up the market.

As to the core technologies, from time to time we make some major new advances. One interesting advance, for example, is a patent-pending technology that we announced recently for hiding users' private data on the Internet — sensitive data such as credit card numbers, health care records, or private cryptographic keys. If the data is stored on a server operated by a service provider, users sometimes demand assurances that not even that service provider can get access to that data. So we developed a new, distributed system technology whereby no one apart from the legitimate owner can learn anything about the private data unless multiple, independent service providers all collude or are all hacked.

The "Holy Grail" of Technology for VeriSign

The market has not to date grown as rapidly as we would like because our services need to be interfaced to a wide range of different applications and the interfacing task takes time and money. Today there's a massive number of applications that could take advantage, for example, of digitally signed transactions, but it requires significant development work on the part of the application vendor to achieve that. What I want to see is a simple, standard, software "plug-and-socket" whereby applications can easily be plugged into a trust services infrastructure like appliances tap the power grid. This infrastructure, however, would serve up e-commerce security services, payment services, legally binding transaction services, etc. Maybe XML presents an opportunity for us to achieve that.

Old Companies Harnessing New Technologies

We're certainly seeing a growing number of companies taking up these solutions and using them within their own business environments. The financial services industry is probably the strongest market sector for us. There are a number of banks and, more recently, online brokerages using our services now. We're also seeing strong take-up in the B2B arena where large enterprises use our technology in automating communications with their own business partners. An example would be Texas Instruments, who have a large secured extranet in place for communicating with their distributor network.

As to developing new technologies in partnership with older world companies, that certainly happens a lot too. We have strategic technology partnerships with many of the big names, in both the technology and non-technology sectors.

Challenges Technology Will Provide From a Security Standpoint

With today's wireless technologies there are some technological limitations that make it impractical to simply extend what we do in the wired world into the wireless world. It would be ideal if you could simply implement on a wireless handset the same functions you do on a browser on the Internet, but bandwidth limitations and resource limitations on the device get in the way. That has forced us to develop some different protocols and different solutions to use on the wireless part of the network. That led to the establishment of the WAP Forum and the WAP standards, which are different in certain respects from the basic

Internet standards. However, security issues arise if you have a different set of protocols in the wireless environment and interface at some gateway point to the Internet protocols used in the wired environment. Because data has to be decrypted and re-encrypted in the gateway, there are potential security exposures at that point. You don't want to expose sensitive data such as a customer's account and PIN in a gateway when banking transactions are going through. That doesn't necessarily invalidate the gateway model—it just means that you have to apply yet another layer of security and have true end-to-end security functions, at least for highly critical data. We're in the process of defining that right now in the WAP forum.

However, the other side to this issue is that we all know that the bandwidths of wireless networks are going to be increasing in a major way over the next couple of years. The capability of the handsets will also increase. So we may get to a point very quickly where the handsets are able to employ exactly the same technologies as on the Internet, and that has enormous attractions. So while we're going to have different flavors of security for wired and wireless Internet access for at least an interim period, we expect the technologies to converge before long.

The Development of Wireless Technologies

Well, there's another problem area that is fundamental to pocket-sized wireless devices and will almost certainly require new technologies—this relates to the user interface. With a wireless handset it's enormously difficult to enter on the keypad a URL like www.yahoo.com. There really have to be some new developments in the way that users get access to sites and it can't be totally resolved by simply navigating menus. I think we'll be seeing some new fundamental changes in the way that businesses are addressed. For example, a more number-based addressing system rather than a character based system. This is one of the areas where we see significant changes being needed and we would anticipate more moves in that direction — which means basic new infrastructures to support such addressing systems.

Exciting New Technologies

One of the more interesting directions that we're seeing emerge is the online peer-to-peer data exchange model as exemplified by Napster and Gnutella, as opposed to the traditional client-server model. It reminds me of the shift in the '70s from mainframe-based computing to minicomputer networking and the early Internet. In both cases, the change implies a massive scaling factor over the interconnection complexity of what went before. Peer-to-peer will drive the demand for specialized infrastructural support services like security services from the enterprise level to the residence or even personal level—in essence, every home desktop becomes a server.

Multimedia is also getting close to the serious takeoff point. Multimedia-related applications will likely be exploding soon, such as multimedia conferencing and interactive multimedia sites, along with voice-based services such as real-time language translation in conjunction with voice over IP.

More Technologies…

The spread of broadband and the extension of the Internet to the wireless world, with Moore's law still in play, means there is enormous room for innovation over the next few years. What we're seeing today is just the start.

Deciding What Types of Technology Becomes Mainstream

The technology needs to fill a major perceived business or consumer need, be easy to sell, distribute, deploy, use, and support, and be priced compellingly. While that may be easy to say, the reality is that attempts to answer this question have filled many books, but we are still in a state where most endeavors fail and few succeed.

100% Security on the Internet

There's no such thing as 100% secure. However, we're very close to the point where any given business can make its own services as secure as it needs to. If you take a particular service provider such as Yahoo or eBay all the basic building blocks are there for them to make their services secure, both for their own protection purposes and for their clients. However, security always has costs and it's ultimately up to an organization to decide if the likely benefits warrant the costs. That's called risk management.

International Security

I'm not sure if international security is necessarily harder. Historically, there certainly have been issues with regard to the ability to use cryptography internationally, but those issues are now going away due to recent relaxations in government regulations. To understand VeriSign's international position, I should first explain how our business lines operate in general, and I'll use our digital authentication services as an example. We provide services through at least three fundamentally different models. One is that we provide services ourselves, acting basically as a trusted third party. For example, when it comes to issuing digital certificates for secure web sites for the public, VeriSign acts as a trusted third-party certification authority in authenticating those sites. The next model is that we enable organizations to be their own certification authority to protect their own communities. We have roughly 1400 enterprise customers in that category now. An example would be a bank like Bank of America. The bank is a certification authority and issues credentials to it's own commercial banking customers, but its operation is supported through back-end services operated by VeriSign. The third model is that we enable large service providers to do everything that we do. This means that they can operate their own trusted third party services and they can sell enabling services to their own corporate customers, operating out of their own secure data centers. Presently, we have 30 to 40 affiliates operating under that model. The majority of these affiliates are international service providers, including BT, KPN Telecom, Telia, and Telefonica, to name just a few. The model supports continued expansion. I see this growing quickly into a massive global network of service providers, offering all of our services, including naming, trusted third-party, and payments services, all under one umbrella of service consistency and quality standards.

Keeping the Company Technologically Nimble

At VeriSign we don't have difficulty in keeping up to speed because we're very closely linked with other technology companies through an enormous number of partnerships. Also, being a young, Silicon Valley-based company, our employees tend to have a natural inclination to keep up-to-date themselves with where technology is heading. Also, we are fortunate that our executive ranks comprise almost exclusively the types of people who understand and track technology developments and evolution closely. To be successful, it is always necessary to have one's own technological innovations but to also leverage external developments when that represents the fastest or most economical path to market. On the research side, we tend to be very market-conscious and rarely expend significant resources in the absence of at least a rough business case supporting an expectation of short or medium-term returns. Awareness of external developments is therefore fundamental at virtually all levels of our company.

Interaction With the Management Team

I am in an interesting position. My strategic technologies unit is separate from the engineering organization where mainstream development is done. I interact directly with the corporate executives, the engineering organization, and the marketing and sales organizations. Getting all these groups on the same page where technology is concerned is really key to my job. I interact directly with our CEO, Stratton Sclavos, a great deal—Stratton has a very strong technological background and is very much a part of shaping the technology vision along with other aspects of the corporate vision.

Building a Technology Team

In my case, which may well be different from other CTO's, I am directly responsible for just a small specialized technology strategy team. The approach is to keep it very lean and mean, and I've been fortunate to find some exceptionally talented individuals that have all the skills needed to fulfill this role. While there is enormous competition for talent today, VeriSign has fared better than many other companies, through building a reputation as one of the best employers in the Internet game.

Acquisition of Network Solutions

The acquisition has opened up an enormous number of new and very interesting opportunities, as they have brought in some technologies that complement our prior technologies perfectly. In particular, Network Solutions has brought in what amounts to a very sophisticated, high volume, high availability directory service that is ubiquitously available across the entire global Internet. This presents a big opportunity for integrating security and trust services into those same points of presence. The Network Solutions acquisition also brought in some interesting technological challenges. The Domain Name System, which is the underlying core of the Network Solutions business inherited from the Federal Government, was inherently not secure. By leveraging the existing VeriSign technologies, we're now able to work together to ensure that the security of the overall Internet core is upgraded to raise the security bar for the whole community. The acquisition also presents a number of opportunities on the marketing side, in terms of being able to package solutions from the two prior businesses. One of my responsibilities has been to help bring these different technologies together under one coherent strategy - one coherent vision across the combined set of companies.

Combining Different Technologies in Light of an Acquisition

Yes, acquisitions can certainly be a challenge. In the case of VeriSign and Network Solutions it's been enormously assisted by the fact that we had very similar corporate cultures, and similar skills and aspirations on the parts of the individuals involved. Therefore we quickly established very close working relationships between peers in the two companies. Given that environment, I think this is more a matter of working systematically through the consolidation issues and decisions. Certainly the whole atmosphere of the combined set of companies is excitement about the opportunities more so than the challenge of realizing them.

Advice for an Internet Startup

I think the key thing is to see results early—see the concept proven in the real market. This means structuring the product rollout plan in such a way that small pieces can be rolled out and proven or disproved in the true market, and then used as a base to be leveraged by the next stage of the plan. It's absolutely essential to be adaptable to rapid changes in the market environment. The market development phase for any new technology is just so dynamic.

Becoming a Leader

The required attributes for different C-level positions vary a lot. However, some attributes are common to all of them. Apart from being worldly and wise, they include willingness to work long hours, to always behave in a way that engenders respect from others, to be a good listener, and to accept that defeats are part of risk-taking and that risk-taking is key to being successful. Related to the last one is thinking energetically and innovatively when the chips are down. Also, the best leaders inherently know that their success is tied to success of those around them—employees, certainly, but also customers and industry partners. And since we don't know today exactly who our customers and partners will be tomorrow, this means building respect widely. It's also beneficial if you have that rare skill of being able to kid yourself that sitting long hours in an aircraft seat is enjoyable.

The Dream Technology

I think we have an enormous opportunity for not just a new economy, but a whole new world, if we can overcome the education and poverty barriers that face such a massive part of the world's population. That's what I'd like to see technology applied to. I think that, given where we are today, the best way to make progress on that is through improved, affordable and robust personal communication systems that not only provide communications capabilities, but also can act as an education platform. So it's a portable device that pulls together all the best of wireless, PDA, Web searching, and interactive learning technologies, with a totally intuitive user interface—and it sells for under $20, and it doesn't break when you sit on it.

Warwick Ford is the VP of Strategic Technologies and Chief Technology Officer at VeriSign, Inc., a leading provider of trust services for e-commerce and Internet applications. He is responsible for VeriSign's technology strategy and the exploration of new product and service opportunities. Dr. Ford is a recognized authority on information security and the application of cryptographic technologies, and led the development of various ISO and Internet standards in these fields. He is co-author of "Secure Electronic Commerce: Building the Infrastructure for Digital Signatures and Encryption." He holds the degrees of B.E. (Queensland), M.E. (Adelaide), and Ph.D. (Toronto).

RON MORITZ

Building Leading Technology

Symantec

CTO

Background

I did not follow a standard path to my present role as an executive at Symantec Corporation. I really have covered a lot of different areas in programming and technology and strategy and, in fact, I even spent a few years in the audit side of the business. I have three degrees from Case Western Reserve University. My B.A. was in mathematics and, upon graduation, I joined a software company as a COBOL programmer writing vertical business processing solutions for the food brokerage industry. When you go from an engineering school to a business software company, something very strange happens. Suddenly, the problems you are solving no longer have numeric input and output. Now you are working with text string manipulation, data base concepts and a complete shift from the engineering background that you were tuned into as an undergraduate.

So it was a truly interesting experience going from my college into the industry. I spent several months at this company writing software for the firm's internal customer and technical support teams. This was 1985 – before you could purchase a customer relations management and bug tracking system from CompUSA or your local Siebel sales rep. What I discovered was that as much as I had enjoyed programming, it wasn't my strength.

Skills Developed

I hit a career inflection point in the summer of 1996 when I joined my second startup. My wife and I were looking at the possibility of living overseas and Israel had emerged as a powerhouse in information security technology. The country is literally flooded with world-class technologists and Internet security startups. I concluded that between my mathematics B.A., my M.B.A., and my M.S. in computer engineering, I was marketable over there. I spent a couple of weeks in intensive interviews and found one individual, Shlomo Touboul, that had a tremendous amount of charisma and energy; we connected immediately. Touboul had recently sold his first company to Intel and had just birthed a new company, Finjan Software, in response to the new and emerging threat and security concerns around

Java. Java was the hot Internet language, one of the hottest topics in 1996, and the idea of building software to secure Java was intriguing. So I joined Touboul at Finjan with the intent on moving to Israel to lead the security strategy. Before I could actually pack my belongings and sell the house, I began working for the startup from my basement in Cleveland, Ohio. Touboul concluded that it was helpful for him to have eyes and ears on the ground in the U.S. and asked me to help him build Finjan's U.S. operations. We hired a U.S. President and a VP of marketing and sales in Silicon Valley and I spent nine months commuting between Cleveland and Santa Clara, California.

The U.S. President, Roberto Medrano, now a GM with Hewlett Packard, became a mentor to me. He turned to me in January, 1997, and said, "We're going to make you an Internet security expert." My response: "What do you mean an expert?" At the time I was quite naïve, still reading the trade press, and believed that the real industry experts included Li Gong, a PhD who, at the time, was leading the Java security team at Sun MicroSystems, Princeton University professor Ed Felten, and Dr. Gary McGraw of RST Corporation (now Cigital).. These were the real experts; how would I compete with these guys? What I learned was that from a marketing perspective a person could become a recognized industry expert. Experts in this industry are like politicians – they're made, nurtured, and coached. If you study up on a particular subject and you find a passion for it, eventually you will get into as much detail as any other expert at that moment in time. In fact, you do start to move in their circles. It's about publicity, it's about getting the exposure, and the marketing side of the business was what suddenly emerged in front of my eyes. I had not been exposed to it before, as I had been isolated in Cleveland.

The company I had been working for — Finjan Software – was so over hyped in the first six to twelve months of its existence that the hype may have even hurt our software development and engineering efforts. We were one of the "glamour children" of 1997 and it was a phenomenal place to be. The rush of being in the middle of the Internet industry, the exposure to people I met along the way, and the interactions I had with the folks in Silicon Valley and Israel gave me a much better education than I had ever received in the isolation of the university. Every day you were living the case study. You were the Harvard Business Review case study.

The Value of Experience

In Silicon Valley you were trying to negotiate deals that from an engineering perspective weren't that clear, but from a marketing and sales perspective were very clear. If we were talking on the phone with the guys at Netscape, (Netscape was in a big high at that point), and we weren't making any progress, we could just hop in the car and drive five miles and suddenly we're at their corporate headquarters. Now your guys and their guys are meeting face-to-face and you've hashed out the issues and in an hour everybody goes home and you move on to the next problem, the next alliance, the next company you want to partner with. I was able to understand and appreciate the networking side of the industry: how communication is key to the success of both companies and how no company could do it alone. That's why the industries' collusion and collaborations were so important. It really broadened my strategic and business skills. Skills that I didn't have coming from the isolation of Cleveland and isolation of the university.

At Finjan Software, I spent a lot of time between 1997 and 1999 commuting between Israel, the US, and also the rest of the world. I had assumed the function of a CTO. I was in a strategic position. I was working across the company without necessarily having direct responsibility for any specific function. The role was like a cross between the manager and the key pinch hitter on the baseball team, coming in to save the game whether it was on a sales call, a business development opportunity, or interacting with customers face to face. Sometimes it was linking companies, two tech companies together and helping set the vision or the opportunity in place so that not only the sales people understood it, but also the engineers. It was becoming a bridge across all aspects of the company, the product management team, the engineering team, the sales team, the marketing team and ultimately the business side outside the company as well.

Important Skills for Gaining Momentum Professionally

What I found over the years, and I think a lot of my experience in audit actually supplemented this, was that the ability to influence outside the scope of my direct reporting hierarchy is a very significant skill. If you don't have that I don't think you can function as a senior manager and certainly not a senior manager in a technical role. If you don't have the ability to bridge communication, both inside your company and outside your company, and by bridge I refer to the idea of bringing two people who don't have the same experience base to a common understanding so they can see the same goal or reach the same or common goal. That's very difficult, and not a lot of people do that well.

There's another concept in my background that has matured over the years and that is the ability to go deep as well as broad. Not many engineers have that ability. I can still sit today opposite any one of the engineers at Symantec and debug code with them. I understand the logic and understand the technical details. I can get into any one of our products as deep as I need to go, but I can also step back broadly and understand the relationship that product has in the market or the opportunity a particular technology has to win along with the other opportunities that are in play at the company. This is extremely important when you look at emerging opportunities or what we call from the old Harvard Business Review case study, a disruptive technology.

When a mature company, not a startup, but a mature company like Symantec (which is generating through it's existing product line over $1 billion of revenue per year), looks at the opportunities that will generate only a million dollars next year, it is very hard to justify an investment. You're naturally inclined to support the innovation that's going to result in very high returns rapidly because your goal is to drive revenue growth. You're not going to drive revenue growth with a million-dollar-a-year application. This is the difficulty a large company faces. When you sit in the innovator's seat and when you sit in a strategic technology seat you have to understand the short-term consequences of the technology investment as well as the long term consequences and the overall effect that technology may have on the company, including the effect on it's reputation and credibility. That is the difference between being a leader and being a follower.

That's what I mean by being able to go broad. Being able to understand much more complexity as you evaluate your technology decisions. Let me bridge you from Finjan to Symantec. I actually came back to the U.S. in the summer of 1999, with the intent of finding or marrying off the company that I had effectively built. Although I wasn't the founder, I was probably one of the most influential people at that company in terms of it's ability to survive over a significant time period, three and a half, four years at that point, and get it to the point where it was. I became one of the most critical people in the sale discussion, the M&A activity that we were pursuing. One of the companies that I had always felt was the right company for Finjan to integrate with was Symantec. We visited with Symantec in September of 1999, had a three hour meeting with the

CEO, John Thompson, CFO, Greg Myers, and the executive VP of Sales, Dana Siebert. I found John Thompson to be a very impressive individual. I concluded that I would love to be able to marry the two companies together as he seemed like he would be a great guy to work for. Apparently he had similar ideas about his interaction with me that I wasn't aware of at the time. We pursued the relationship for about three months.

At the end of the three months in November of 1999, the Symantec folks who had been performing the due diligence sat down with me and basically said, "don't call us, we'll call you." Obviously, this was extremely disappointing, but the very next day I received a personal e-mail from John Thompson. Basically he outlined a whole list of very interesting opportunities at Symantec, and said, I'm sorry the deal didn't work out for Symantec and — but I think Symantec would benefit from having you on board. Being the salesman that he was, he had really hit on all the passion points that I had. One of these passion points was building a science up capability within Symantec. Academia is an aspect of my background I've always enjoyed. I would have like to have pursued the PhD. I didn't, but Symantec was offering me an opportunity to surround myself with a bunch of PhD's and pursue Internet security research.

Thompson came from 28 years at IBM, 28 years of understanding the value that Watson Labs provided to IBM. While it wasn't necessarily his intent to create a lab that would research anything and everything, the idea was that a lab that would incubate wild ideas in the world of Internet security could help Symantec develop leading edge products in two or three or four years out. If you actually look at old research publications, you will find that many ideas discussed by research scientists in 1996 have appeared in products by 2000. On an on-going basis, I review the academic pubs across four-year periods. This year I'll go back to the academic publications from 1996 and correlate a particular innovation from 1996 with a particular company championing that idea today; the four year cycle is pretty consistent. So, the idea is if we engage in some of that raw research, raw science now, we could deliver products that are really interesting to the market in three to four years. It's not just about buying startups for a larger company, it's about making technology investments that are long term. Very few companies do so successfully, and one of the only universities that has a successful technology transfer function is MIT, the others don't do it that well at all.

Emerging Technologies

There has been a lot of discussion on wireless and what wireless means. The wireless Internet has surprised me; that is, how fast it's becoming a point of interest or a focal point for many of the new and emerging ideas. I'm looking at the opportunities Symantec has, specifically in this world of wireless, but what it's really shown me is that many of the possibilities have yet to be imagined. Let me give you an example. When we talk about the next generation of a home appliance, we talk about the sexy aspects of it. That is, the refrigerator with the bar code scanner so that every time you take the carton of milk out and put it back in, your usage is tracked. A just-in-time inventory management system for your groceries providing automatic computation of order requirements. If you're a subscriber to a grocery delivery service such as Web Van then your refrigerator can submit your grocery list automatically to Web Van and the next day the just-in-time system is working on your behalf and you've got a whole bunch of fresh milk delivered to your door. That's the science fiction stuff, that's the sexy, glamorous stuff that everybody thinks about and says, yeah, yeah, when that happens that'll be great. But you know, there's a certain reality that says, do we really need that? There is a reality that looks at a different type of appliance and says, hey, there's something here that could in fact be very interesting, and that other appliance might be say, a washing machine. A washing machine is a mechanical device, it's got bearings that allow the drum to spin and many other hard components. Those components fail and there's a known mean time before failure. The device itself can do a pretty good job of self-monitoring. That is, a health check. The components themselves know when they've deteriorated so why not add a simply measurement device, a sensor, that could notify someone when a component is about to fail? Automobile disc brakes have sensors that measure the disc thickness and notify the driver when the disc is too thin. Similar sensors could be added to the washing machine. Washers know when they're about to break and if you put the right sensors in place, you can imagine the day that you walk into the house with your new Maytag and it registers itself with a local service provider. That is, when you plug the washer into the wall where your electricity provider has updated the electrical infrastructure to support Internet communication, the machine uses that data communication path to register itself with a local service provider. At a future date, the service provider receives a message from one of the washer's sensors that has been triggered regarding a potential component failure. The sensor may even be able to estimate the anticipated fail date, for example, in ten more cycles or about two more weeks. The provider can call you and say, "your washing machine has indicated that it's probably going to fail in two weeks, when would you like a service call?" Emergency service is quite expensive. The reason you pay so much for such service is that it's very tough for a maintenance organization to maintain the staff and the inventory that they need to service your machine. If they know that a particular piece of hardware is about to fail, their just-in-time inventory system can also operate on your behalf. Since the service provider need not stock his shelves with extra parts in anticipation of your machine's failure and since he can schedule his technicians in advance, his costs are reduced and so are yours.

As you can see, there are many down-stream services, products and systems that are affected by communication from devices. It's much less glamorous than the idea that your refrigerator's going to order your food for you, but I think it's much more real. I talk about the day when I can get on the plane in San Francisco, fly to Washington, D.C., get off the plane and go to the conference where I need to do a presentation with just my cell phone in my hand. I think about the day when the cell phone will communicate with the NEC or Proxima projector via short band wireless like Blue Tooth. And, on the other side of the cell phone, the traditional cell phone network will tie me back over the wireless Internet to Symantec servers that will host my

PowerPoint presentation. The PowerPoint presentation will then be transmitted over this wireless Internet to my cell phone. From my cell phone it would go to the projector and I would control the whole presentation using my cell phone. That to me is a very real direction for the industry.

Another concept is intermediary computing devices. When I think about this area I think about the security infrastructure that has to be in place for it to be possible. I think about this world in the sense of, how do I protect the data, how do I protect the individual, how do I protect the privacy of transactions? What are the services I need to provide so that less secure applications can sit on top of me? A really good example of this is a system like the Air Traffic control system. This is supposed to be much more robust, much more fault tolerant system than an application system like an airline reservation system. When the airline reservation system fails, customers become irate and there is a loss of confidence. At worst they may get into fistfights. However, if they get into fistfights, that is a very different effect than if the air traffic control system had a failure and planes start to collide or planes end up in the wrong place. Then you're putting lives at risk.

When you look to build Internet infrastructure, you need to have solid hardware that's dependable, reliable, and that's fault tolerant and you build applications on top of that hardware. Those applications will be reliable. The routers and the other data communication hardware that we've developed over the last decade have had to be solid. But on top of that hardware there is another layer, the security layer, that has to be just as hard, just as robust, just as fault tolerant as any of the other layers that it relies on, the hardware layers. What inevitably happens in the rush to develop new business applications is that we create systems that are insecure. At the very least, such systems must sit on a secure infrastructure, a secure base so that they can succeed. So when we're talking about next generation e-commerce, next generation B2B applications, there has to be a secure infrastructure.

Perhaps the following story can help frame how critical it is for that secure infrastructures emerge. Symantec and Yahoo! engaged in an interesting relationship several months ago whereby all attachments emailed to Yahoo! users or emailed by Yahoo! users are automatically scanned by Symantec anti-virus scanners. To celebrate that relationship, the Yahoo! folks invited the Symantec folks down to the new San Francisco Giants Stadium where Yahoo! maintains a very nice suite. I was a little reluctant as I'm not a huge baseball fan. What convinced me to attend was that David Filo, Yahoo!'s CTO and co-founder, was going to be there. Perhaps he and I would find some common ground and have an engaging and stimulating conversation. Basically, we spent 9 innings trying to find something to talk about and it was very telling to me. Here you have one of the visionaries of the Internet, pushing the fringe of the technology and not wanting to be restrained by any of the security principles that guide my visions. A secure infrastructure is not in Filo's vocabulary. Every single time he suggested a new opportunity, my mind was trying to figure out how to protect that. Every time I introduced the principles of protection, security, safety and privacy his response was, "How does that affect my ability to get to that next inflection point for the Internet? How do I get to that next technological hurdle if I have to worry about those security issues?" It was very telling and helped me understand why Yahoo! did not pursue an anti-virus service strategy sooner. Protecting email users from viruses was not on their radar even though the virus threat had been around for well over a decade.

This was a very interesting discovery. The art that I had been practicing led me to conclude that you can't build these next generation technologies without thinking of security while the direction that Filo and some of other Internet visionaries are pushing helps carry us to the fringe of the Internet. They can't be restrained by thinking about the security problems as we move to the next level. So it was a very interesting night, I don't think we really ever got to any common ground but it really helped me understand where some of the resistance is to the ideas and the concerns that I have.

The Playing out of Java and XML

These technologies and others that define the Internet have become such a focal point that we have no opportunity to do anything but continue to deploy them. I believe that we will slowly see a migration from platforms like Windows, which have been the main staple of our computing for so many years, to platforms like Palm, embedded Java, and Symbian EPOC. That migration will happen slowly but it will happen. The reason it can happen is that the abstraction between the technical challenges of interaction between various flavors of operating systems and the individuals who are using them, most of whom are not literate in those operating systems, will happen through things like the Web browser. As more significant applications are developed around the browser (possible because of the emergence of extended capabilities of the browser application), the need to actually understand the operating system becomes less and less. Platforms like Linux and Solaris and AIX and even OS/2 are much more stable than the Windows environment. So, if you abstract the operating system from the applications such as word processing, the user begins to worry less about whether he's running on a Mac versus a PC. The prevailing wisdom is that you can do that through a Web interface or browser and, if you succeed, then the users can become just as comfortable in that environment, say Linux, as they may be in any other, more familiar environment like Windows. Java, XML, and other languages of the Web suggest that this is possible; as such, they cannot be ignored.

That's a different way of thinking. It also means that Windows has to be recognized for what it is: an operating system but not the defacto operating platform of the future. From a Symantec perspective, the many operating system flavors pose a challenge in that each requires expertise. Maintaining a group of experts for the Windows operating system and operating environment is critical but just as critical is our need to maintain expertise in several other key operating platforms. This is a new challenge, and much more so to a company so closely linked with one particular operating platform.

What I am suggesting is that corporations must evolve and their applications must also migrate to other platforms.

That's Microsoft's dot net strategy. Forget the operating platform. Port the key office apps to the network so that any device – personal data assistant, mobile phone, laptop, desktop, can use the same service. Those are big challenges and those are possible because of some extensions to the Web that have emerged. We're seeing basic services and capabilities overloaded and extended and used in ways that allow the browser, the Web interface to be more than it's ever been.

Skills in Demand

This is a really interesting challenge for a CTO. I'm an old engineer at heart, I still have the passion for the creation and innovation and I really like to get into the details when I can. Ultimately, the CTO's role is much broader. It's not just about innovation here. I have a whole bunch of sustaining technologies, what I call infrastructure technologies that my groups are responsible for. The challenge is to find the balance between new engineers coming into Symantec, (by new I don't necessarily mean lacking in experience, I simply mean outsiders), and a very deep and experienced team which has been around for many, many years at Symantec.

There are people here very senior, very key individuals with nine, ten, eleven years tenure, unprecedented in this industry. These people have gained a certain expertise that you don't want to lose but you need to bring in more people to grow your business and to focus your efforts. Good engineers are very hard to find in the industry and the need to maintain a comfort zone for your existing teams is important so they don't feel that they're being moved aside. You have to bring in the new people to energize the company in a direction that it hasn't gone before. A very clear example is Symantec, a company built on a foundation of Windows, which is interested in the enterprise market. In this market it is the Unix operating system (I'm using a generic terminology to cover the gambit of Solaris, Linux and others), that is very important. As the company shifts from the consumer and the retail box space to an enterprise focus, its core engineering capability also changes. That makes the older engineers, whose experience is rooted somewhere else, a little uncomfortable.

So, the challenge I have is finding the balance between the skills of the existing employees and the skills needed to move the company forward from the outside. I can highlight some of those skills today. In case you haven't been paying attention, there is a shortage of engineers these days. And, when you are fortunate enough to get them, they are very finicky. It's almost like working with artists. If you can gather enough momentum and herd them in the same direction, then they're a powerful bunch. But because they're like artists, each one has individual ideas, each one has their own thoughts and in many ways they're all over the place. It is very hard to keep them focused, happy, and energized. Keeping their collective eyes on the same goal is a very significant challenge from a management perspective.

Becoming a Leader

One of the key roles of the CTO is to provide the technical vision to compliment the business vision, setting the tone and direction for the company's technologies. Leadership, in this context, comes from being able to set the technical course and from being able to define what the company's products and technologies might look like in two, three, or more years. That is not a default skill.

More generically, a successful leader is one whose projects and progress can be measured using the right set of metrics. CTO accountability is a mystery to most executives. Consequently, the only accountability that makes sense is one where performance can be measured against a plan. That is, attaining milestones, introducing new products, reducing costs, reducing uncertainty, and acting razor-sharp on results whether they suggest projects should be terminated or funded. So to be a leader, whether a technical leader of an R&D group or a leader of a business unit, one must be able to set the expectations and frame the questions. But that's only part of the story. A true leader is one that can get his people as excited as he, regardless of the project and often times in the face of adversity. This allows the people to feel that they are an important part of the enterprise instead of adjuncts to it. That adds up to productivity and productivity allows leaders to reach the goals they set.

Advice for Other Companies

Understand that my focus is security so my advice comes skewed by my art. Basically, the advice I can offer is that you can no longer avoid the security challenges. Computer security has become, and if it hasn't in your organization it will soon, a boardroom issue. There an understanding that the boards have a fiduciary responsibility, but they also have a responsibility to protect the information assets of the organization. In order to protect those information assets, you need computer and Internet security to be a top-level priority. Most of the hands-on level folks who are going to be responsible for building and implementing security services will simply try to brush it off as one more thing on a long list of things. The guys who are responsible for making the technology purchase decision and the technology implementation work are not happy about all the emerging security services that they have to deal with. Often times you'll hear these folks looking for a way to defer this project or that for another year. At the same time the line of business guys are running a million miles an hour toward new opportunities. More often than not, the new opportunities help set the technology strategy for the company. They will publish web sites that are insecure. They will build services that tie into back end information systems and into the back end databases. They are simply not considering the security ramifications and the guys inside the corporation with security responsibility are simply left behind.

If you don't tie your business strategy with a security strategy, you will be in trouble as we've seen with many companies. A good example is CD Universe or CD Online or one of those music-for-sale-online shops. That company hadn't taken the steps to protect their Web transaction processing system and customer database. They were attacked earlier this year and the credit card information for their customer base was released and published. As a result, they lost their business. That's a good dot.com example where a company rushed – at Internet speed – to get online.

You can also see similar problems for much larger companies like CitiGroup or CitiBank. In the early '90's, Eastern European hackers (I believe Russian) attacked CitiBank. At the time, $10 million was siphoned out of the bank. All but $400,000 or $500,000 was recovered and at that point the bank felt it was important to share that information so that others in the industry would benefit from the knowledge about that attack. They shared how the bank's security guys were able to track down the money, how they were able to track down the attackers and how they were able to prosecute. However, in revealing that, they got a big hit against their brand. Subsequently, few organizations have been willing to share information about their attacks, at least not publicly. This hesitation to share attack information has obviously hurt our ability to focus on ways to stop new attacks.

I think the bottom line for a company is, pay attention to the security requirements of your new applications and make sure the security risk is managed opposite the opportunity. Otherwise, the impact on your brand may be quite serious.

The Dream Technology

I've always been enamored by the Star Trek sci-fi where you beam people and things from point-to-point. With the amount of travel that I do, such technology would be really effective. I don't know whether my attraction to this transport technology is consistent with that of others in my industry, but it's certainly a glamorous concept.

We've pretty much covered many of the other ideas that our sci-fi authors have proposed but not the idea of being able to move from point to point instantaneously. You have no idea how many airports I visit each year. Let's just say this year I will have attained the top-level rankings on two airlines.

Advice for a new Internet Company

Focus on the technology, not on the hype. Be careful about what the marketers in your organization are going to do to you. I've seen the effect of hype on a company. At Finjan we were energized up and down the organization, the engineers, the production people, everybody, but it was also so distracting to us that we didn't run our race as effectively as we could have. What it comes down to is, don't buy your own hype. I don't think that's a new principle. I think we've heard about it from other people as well.

And the other piece of advice is to focus on revenue. You may be excited about the technology but, at the end of the NASDQ day, it's all about revenue growth. If you want to invent, join academia. If you want to be part of a startup then you better know how you're going to pay the bills. Finally, read Jerry Kaplan's book, "Start Up."

Exciting Aspects of Technology Life

My answer will be very specific to Symantec. I've tried to keep the other responses broad, but I'm actually practicing my craft at Symantec today for a particular reason. Joining Symantec was an inflection point in my career. I have a passion for small companies, for startups. To join an organization that employs 2,800, an 18 year old company that's been there and done that, a company with a very broad portfolio of software, was not exactly consistent with my passion. But it is great to have an opportunity to sit next to a CEO with a game-changing business vision and to help him tune the technical vision for this company. I can't imagine a better opportunity than that, especially since the technical vision that is demanded is precisely where I've tuned my career to be. I want Internet security to be a key focal point for the whole dot.com industry. Where am I going get a chance to do that more effectively: at a company that is going to be the Internet security Company or at a company that's going to be a niche player in the Internet security space? I can spend as many years as I want with startups, but the place where I have the most influence not only from a company perspective, but also from an industry wide perspective is Symantec. It just doesn't get any more exciting!

Ron Moritz is Senior Vice President and Chief Technical Officer at Symantec Corp., where he leads the Symantec Research Institute, leveraging the company's investment in research and development. An Internet security expert with 16 years' technical experience, Ron is responsible for building on Symantec's existing research infrastructure to address the growing number of security issues facing the industry. An established expert in Internet security, Ron presently serves as a member of the U.S. delegation to the

G8 conference on safety and security in cyberspace. He has published and presented numerous papers on a variety of security topics including mobile code security, Web browser security, security and electronic commerce, and computer ethics and privacy. In addition, he is one of a select group of CISSPs, information systems security professionals certified by the International Information Systems Security Certification Consortium. Ron attended Case Western Reserve University in Cleveland where he earned a master's degrees in engineering and business administration and his bachelor's degree in mathematics.

DERMOT MCCORMACK

The Business Sense Behind Technology
Flooz.com
CTO & Co-Founder

Background

Basically I'm the CTO and founding technologist here at Flooz. I started the company coming on about 2 years ago with Spencer Waxman and Robert Levitan. Robert and I used to work at iVillage.com. He was one of the founders and I was one of the early people there and we helped build the sponsorship sales group. Before that I was founder and CTO of a company called inW@re technologies, which was founded around the end of 1994. That was my first Web company, but a lot of people didn't know what the Web was back then and didn't know what to do with it. I was trying to pitch Web sites when nobody knew what I was talking about. So we wound up doing a lot of intranet, extranet work at the time. People didn't see the value in Web sites, but they saw the value in internal networking and they saw the value in the type of efficiencies an intranet and extranet could provide. We also consulted for banks and other big institutions about what this new future might bring. When I told them the amount of e-commerce that would take place in the next 5 years, they didn't believe me (I almost didn't believe me!). I ended up selling that company and they relocated to Ireland, where I'm originally from. I then moved on to iVillage. Part of the reason I moved to iVillage was that I had a lot of the technology parts and wanted to expand my horizons into business and sales. My initial background is engineering with degrees from Trinity College and Dublin Institute of Technology. I specialized in automation engineering. I worked in a few different places such as France and Ireland. I then came to the United States and worked in networking and basically saw the Web come along and thought this was going to be a very interesting sort of revolution. I guess everybody realizes it now, but not many people did then. I liked it because it allowed me to use my left brain with my right brain. I used my engineering skills and combined them with my creative skills and then later with my business skills. I think one of the most interesting things about this whole new era is you're not just required to be a technologist — you have to be all things to all people. That's why I became interested in it. I decided to go somewhere like iVillage because I wanted to learn a little bit about the economics of the Web and how some of the sales and marketing type functions worked. I had to learn what these new business models and entrepreneurs were trying to create around this new technology. To me, it seemed like the right place to go and it turned out that it was. In 1995 and 1996 there were a lot of the interesting business models starting to emerge. For example, you had millions of computers connected around the globe and people were downloading pictures of Cindy Crawford, but how do you make money from all this? Particularly in the early days companies took advertising and magazine type content models and just put them on the Web. It ultimately didn't work and people realized that this "new economy" required more complex business models, not just digital versions of previous ones and that these models would have to have old economy values like making profits. That's why I liked the concept of Flooz. It wasn't your standard dot.com business model, it was something different, something that hadn't been done before. People hadn't actually sent gifts to each other via e-mail before. People hadn't provided each other with an electronic value that you could spend in stores online without leaving a desktop, so on and so forth. We solved a problem and I think if you look at a lot of the successful Internet companies so far, whether it is the Amazon's or the Yahoo's, and it's a matter of opinion whether you think they're successful. I think they are. They actually solved a problem, they took something in the real world that didn't exist and created it online. Both of them are fairly similar. Amazon didn't start off by calling itself Books.com and Yahoo didn't call themselves SearchEngine.com. They put a brandable name on a killer-app service like selling books or searching the web and wrapped everything they could around it. The fact that Yahoo hosts fantasy football leagues and sells airline tickets, four or five years ago this was a very difficult progression to see. You probably remember Amazon when they started up, if I had told you then that in five years they will sell cars and lawn furniture, you may have said, I don't see it. In a way, that's why I like Flooz, I think it is well positioned to take advantage of the next few progressions of these mediums just Amazon and Yahoo took advantage of the previous ones.

The Advance of Technology

I believe the stage I just described was the evangelistic stage. You said, this is great, this is going to happen, it's a revolution and people were saying, get the @##$% out of my office. The second one that came along was community. I think the mainstream started to figure out what academics and techies had known for a while, that there was a lot of value in an electronic community. People figured out how to use message boards, chat rooms and content and how to communicate with each other. The third wave was pretty much e-commerce. Buying and selling through the Web, consumers found more flexible ways of doing things and businesses found ways to become more efficient. Now, I believe, we're heading into a fourth phase. I think the fourth phase actually starts to combine all of these elements and starts to take advantage of the next wave of technology, which is probably going to include broadband and wireless. We went through those different phases from pure utility to content and community to e-commerce. The next is how do you tie all this into the real world? Electronic commerce is not going away nor are bricks and mortar businesses. How do you create synergies between those two worlds? How do you combine these new synergies with new technologies and actually make money? "eBusiness"and "eCommerce" WILL go away and will be replaced once more by plain 'ol Business and Commerce. This is what I believe the future is all about.

Areas of Expertise

When I began my career, I specialized in automation engineering. But now, I would have to say that my area of technical expertise would revolve around all that is required in building an electronic business infrastructure from the ground up. I think what I've done most in my career and what has become my area of expertise is building successful, scalable systems from the ground up and understanding how each part integrates with each other. I've been very successful at doing all this in extremely fast Internet time, which is actually the biggest challenge of all. Almost anyone can build great systems given unlimited time and resources. I think the challenge is doing that with three people in four months, and with $500. Flooz so far is only 22 months old and we've gone from three people to 93. In May we had over a million e-visitors to our site and we basically just spent 19-20 months building that from almost nothing.

How Broadband Will Affect Building Scalable Web Sites

It's analogous to highway design. What I mean by that is if you're designing a highway and you add two lanes, then initially it's going to ease the pressure because you'll have fewer traffic jams. In the long term however, you've just increased the amount of traffic on the highway, and that will increase traffic jams again. That's sort of how I see broadband affecting the whole thing. The amount of bandwidth you have will always be filled up by something.

I think the people building web sites in the first few years had to build a lot of things from scratch – that will change over the next few years. There will be a lot more tools available to build these electronic infrastructures going forward.

My overall feeling is that all the technology will start to settle down and it won't be such a monumental thing. When was the last time you heard of someone that built a TV station from scratch? And if they did, how much of a hassle was it to build? They could've just gone and bought everything off the shelf. When was the last time you heard 50,000 viewers turned on their TV and the TV station crashed? It doesn't happen anymore. The same thing with radio, it doesn't crash, but it probably did for the first five years of radio.

Major Technologies Used Today and Their Role in the Future

Well, on the front end right now HTML is the standard presentation protocol of the Web. I think that's going to be replaced, I can pretty much bet the farm it's going to be replaced pretty much everywhere across the board by XML. XML is a much more flexible and powerful way of representing and exchanging data. One of the trends I'm starting to see now is not just the ability for companies to present data to a consumer or to a partner via a website but the ability to exchange data with many different partners through many different systems and channels. To do that you need a common language and XML seems to be that common language. We know users are already connected via the Internet, but backend systems are starting to be interconnected so the data can flow more efficiently. If you're in business and you need to get data back from your merchant partners or if you need to get billing information to your transactional partners, or if you're a bank and you need to get procurement data to your suppliers, then you'll need to have server to server communication. XML is a great language for that. I think XML is something that everybody should be exploring and using. With the proliferation of new presentation devices like wireless phones, PDAs, TVs etc. , you're not going to be designing just for PC's. Designing something in HTML and fitting it in the other protocols just won't work, your system will have just one output (most likely in something like XML) and each presentation platform will have it's own specific style templates. WAP will probably be replaced by more broadband friendly protocols like Japan's iMode as well.

On the server front, I think you'll see a large increase in cluster- based Linux systems as corporate America starts to adopt this technology. On the middle tier level I think you'll see an increase in Java, Enterprise Java Bean type solutions. If you're a business with aspirations of e-infrastructure, you're not just going to be serving data to an HTML browser. You will have to inter-operate with many different types of legacy platforms and transactional systems. And this inter-operability will be made a lot easier with middleware and Orb technologies like EJB and front end technologies like XML.

On the database front I don't see Oracle going away anytime soon. They seem to have a lot of the e-business sewn up, but it's going to be an interesting battle over the next couple of years to see what platforms companies chose "e-enable" their business. The main threat to a company like Oracle may come from the growing tendency for companies to outsource applications and infrastructure to ASPs.

Resources in High Demand

It's really across the board. If you come across good Oracle DBA right now you should hold onto them. Also finding great network engineers is probably one of the harder things to do right now, someone who really understands networking. This role will only become more important as the networks get bigger and the traffic on these networks gets greater.

The availability of good Java programmers should increase over the next few years but right now they are still a scarce and expensive resource.

In following the earlier point about inter-operability, another trend you'll probably see is a demand for technologists and programmers that have experience with both newer technologies and legacy systems. They understand what a mature system looks like and they are familiar with the tools to build a new system. Candidates with this skill set will become extremely valuable.

Developments in the Wireless Area

Well, first of all, I think that it's an amazingly interesting area. I think as I said to you before I was around in 1994 pitching this thing called the Web that we all know is out there now, but people looked at me blankly. I see something a little different in the wireless arena. I was out on a recent press tour with a project that we were doing with another company and people tended to get the wireless Web a lot quicker than they got the first Web. I think there is a feeling in some quarters that they completely missed the boat on the Web and they don't want this to be the case with wireless. There's almost a hunger for applications now. People get the wireless Web and they want to be a part of it, but there's nothing really out there that they can put their teeth into. Whereas the first time around there was stuff out there, but people just didn't want to know. But with wireless, it will come, and it will come faster and with more ferocity that the first PC-based wave. The strides in wireless technology and adoption have been much, much greater in Europe and Asia than the US. But I believe the US will catch up and even surpass Europe and Asia in the next few years.

By 2003, according to one report I read, we will have a situation where more people will access the global Internet via handheld devices than traditional PCs. If you look at Europe right now almost everybody has a cell phone, or a mobile phone as they call it. What's huge in Europe right now is SMS messaging, short messaging service. Basically in most of the US right now you can only get one-way SMS, which means you can only get a text message to your phone. You can't send one from your phone, but with

GSM, which is the standard in most of Europe, you can actually send a message and receive one from your phone. Somewhere in the region of a billion SMS messages were sent in Europe alone last month. Now you can't ignore that.

What people are doing is using it almost like Instant Messenger on a cell phone. A lot of kids are using it especially. Everybody's using it, because it's actually cheaper to send a SMS message than to make a phone call. So you can send restaurant directions to a friend and the time you'll be there via SMS, that sort of thing.

The first phase of the wireless Web will be information-driven like the PC-based Web that we just talked about-information like your stocks, your weather and all that sort of stuff. Then the next part is community oriented where you can actually send messages and communicate in different ways. Then comes commerce. There will be location-based services so people will know where you are based on your cell phone location. In theory, you can walk by the Gap and if you signed up online with their promotion, then they can beep your phone as you walk by and tell you it's 10% off T-shirts right now. And when you start to think about stuff like that, it starts opening up all sorts of possibilities.

The Web was supposed to be the ultimate opportunity for one to one marketing. Now just think about the possibilities of the Gap being able to beep you a message as you walk by their store, or CitiBank being able to send your phone a bank balance when you walk by an ATM. Or every time you switch on your cell phone you see Pepsi's logo. If you talk to people like Nokia, they don't call their devices phones, they call them life style devices, and I think that's a clue as to where this area is going.

I'm on the board of a European company called It's Mobile who are based in Dublin, and they're launching a program where you can find a particular movie, purchase the ticket from your phone, and pick it up from the kiosk at the cinema. The twist is, it then gets billed to your cell phone bill. They have another one that involves paying for parking meters. Instead of fumbling for change like one normally has to do, you just dial the number of the parking meter and the meter is activated. My favorite part of this application is if you go for lunch and your parking meter goes down to zero, it beeps you with an SMS message that says, "do you want to recharge it for another half an hour?" This is nothing fancy, nothing crazy, but just a simple little application that people actually want and desire. That's why I like it.

Take Flooz, for example, say it's your Auntie Bertha's birthday and you have no idea what she wants even if you had the time to go get it for her, which you don't. Flooz is the perfect application. If she has an email address you go click, click, click and she gets here gift – one that she gets to choose. This metaphor also extends to the wireless world as you may not get the urge as you're sitting in JFK to buy lawn furniture on your cell phone, but you may get the urge to send your cousin a birthday gift because you forgot his birthday. That's the concept of Flooz. If you tie together applications like Flooz and applications like buying cinema tickets and paying for parking meters, you start moving into transactional areas, so you've gone from just content to some sort of messaging, and now you're going to transactions. Now you're actually using this wireless thing, not for something that the phone company thinks is a nice marketing gimmick, but something that you could actually use every day and actually makes a difference to you. I'm sure you use the Web and the stuff you do now is almost part of your daily routine. You go in, you check your weather, you check your stocks, you go and do whatever you do, but the four or five things that you do now are part of your ritual. It wasn't like that five years ago. Five years before that people didn't use ATM cards so probably in two years time you'll use your cell phone for four things that you don't use it for now. It'll almost be impossible to pry it away from your hand because who knows, you'll probably set off your car alarm with your cell phone. You'll use it as a method of getting money out of an ATM machine, maybe you won't even need an ATM machine, maybe you'll just go into a store, dial up a number and it'll load digital credits like Flooz onto your cell phone. You'll beam them to the point of sale terminal in the grocery store. That's exactly what will happen, then you'll go up to a Coke machine and you'll beam the little signal over to the Coke machine and it will dispense your Coke. Those are my views on wireless. It will start off slowly, but it will see hockey stick growth and in two/three years. People will actually use cell phones for basically four or five things that they don't use them for now.

Emerging Technologies

Well, technical innovation and adoption continues to grow at staggering rates. A lot has happened in the last 50 years, the microchip, the PC and most recently the arrival of the World Wide Web.

The Web is basically the interconnection of a lot of different computers and hardware across the planet to a global network. And it's not just PCs and servers anymore,

I now have small powerful handheld, pervasive devices connecting to this global network

What is all this adding up to? It's starting to be about how information flows, whether it's community, or transactions, or content. It's all information at the end of the day. It's the information revolution that has happened. With wireless devices and wireless connectivity you can actually have information with you all the time. Couple that with increased band width so that really means the flow of information is fairly endless. You're not sitting there waiting for everything. The information can be multi-media, it can be pictures and sounds and movies, not just text.

The information revolution continues and then another big part of it is interactive television which I think is going to be huge. If you combine all those elements together, you'll basically get a society that has instant access and Internet activity at all places on the planet. Entertainment will be a big part of this interactive environment.

You will probably interact with movies, and the tough thing for people like TV Networks or Movie Studios today is that they don't know what an interactive movie or sitcom are really going to be. How do you watch Seinfeld and interact with it? People have no idea how to weave narrative into interactivity, but there's one set of people that do and this became clear to me recently when I was watching someone playing Tomb Raider. The last time I was into video games was when Space Invaders was popular, I kind of missed the whole new 3-D shoot-em up games like Doom and Quake. So it dawned on me as I watched this person play the protagonist in the game, running and jumping and interacting that the people who had made this game had figured out how to make a story interactive and how to weave the "viewer" into that story.

I then realized that these game developers will probably be the new directors in an era if interactive Hollywood has any sense. Someone said to me, well I don't think people want to interact. They want to be entertained. But if I'm 12 years old today and I've been interacting with Sega and Sony since I was 6 years old then that's all I know how to do. So when I'm 18, I want to interact because it's too passive and boring for me not to interact with something. It's the same with my generation who quickly got used to the 3-cut per second editing and 3 minute attention span videos served up relentlessly by MTV. We're now very used to that technique and to us they seem normal.

People who grew up interacting since they were 5 or 6 years old with these things are going to want different entertainment. I think sitcoms and movies will be interactive. If you're a 16-year-old game developer right now I'd probably buy stock in you because you'll probably be a big director in Hollywood in ten years. So that poses the question, what's a big part of what we do in life? It's how we get entertained. A big part of that is the technology that's come along, the evolutionary change and how people can deal with it and multi-task. Kids today can do five different things electronically whereas I can just do two, my mom can only do one, and her mom can't do anything with an electronic device. I think you hear the word convergence a lot and it's not just converging a PC or TV or converging this with that. I think it's converging five different types of things into one with the ultimate outcome that information will be extremely pervasive and it will be interactive. What that actually means in society, I have no idea. If I do, I'll let you know.

Advice for Internet Startups

Giving advice like this reminds me of when people ask Bono of U2 for example, what advice he would give to young, new rock bands coming up. Bono says, well my advice would be get some great songs, don't worry about the image and don't worry about everything else. Get some great songs and practice and I think you might have a shot. I give similar advice, get some great songs and practice. What I mean by that is, if you're an Internet startup company or any startup company at this stage of the game, (I don't even like using Internet startup anymore), I think we're getting out of that stage. It's like the first days of telephones, everyone was a telephone startup, but after a while people just used telephones as a business tool. I think ultimately, the Internet will just be a business tool. Not every single company that uses it will be called an Internet company.

The advice I would give to new technological businesses that are starting up would be to make sure you get the songs right, make sure what you're doing actually makes sense from a business perspective, and make sure that you know you can actually see yourself having some concept of making money some day. Don't just get involved with technology for technology's sake. Don't get involved in something that ultimately you're not going to be able to execute against. Make sure whatever it is, is really interesting. You'll probably convince somebody to give you money, not so much in this environment anymore, but there was a certain period of time where there was a feeling that if you had an idea and an Internet connection you could make a billion dollars out of it. Now it's more about actually something that makes sense, that makes money. The first thing I would tell you is to focus on your business plan, not your technology. If your business plan makes sense, then focus on your technology. Make sure at the very beginning that you get very smart and dedicated people around you. One thing that all successful small companies that grow into big companies have in common, regardless if it's in biotech or technology or the restaurant business, is that there are some passionate people at the helm. One thing that Venture Capitalists have always looked for is obviously a solid business plan, but they look for passion. They look for a degree of passion, but if I give you $2 million and I give this other guy $2 million who has the fire in his eyeballs, who will go and sell his shoes to make this thing work? I'll always go with the second guy because there are just some people that won't lose. They'll just be more tenacious.

So hire some smart people, have a good business plan, believe in what you're doing and have passion. Another thing is to have some fun. Don't just do it because you want to make money. Don't just do it because you think it's the right place to be. Do it because you think it's cool and you like it. You're going to be spending 14 hours a day for the next four years doing it so you better enjoy it. If you don't believe that, then don't do it, go do something else, life's too short.

Acquiring the Technology Skills to Succeed

The truth of the matter is that a lot of the technologists in the Web business didn't go to college to become technologists. They went to school to become economists and environmental majors and just found this area sort of interesting. Of the people who work for me probably half of them have computer science degrees and the other ones don't.

My advice would be, if you're post college, identify an area that you think you want to focus on. Go out and read some books, ask some questions, call up some people you know in the business, find out what it is that you want to do. If you think it's writing Java programs for Middle Ware, then do that. If you think it's something else, then do that. If you don't know, then my two prongs of advice have the same ending. They are to get yourself a job in a company that you like and if they won't give you a job as a programmer, get a job sweeping the floor. Get a job at anything. That's my advice to anybody. Don't sit around, don't spend $50,000 on classes, go get yourself a job and get yourself a job as anything, anything that they'll hire you for, tell them you'll even work for free.

I have probably two or three guys that did that and they have followed me to my second and third companies because they know what they're doing. They said, look I just really want to learn this stuff. I don't know a lot about it, but I'm willing to do anything to get there and people actually admire that. Everybody in these companies needs energetic people around them. Everybody has 50 million things to do, so my advice would be if you're really into it and you like it and you want to do it, pick a company you like and go and see if you can get a job. If you are good you'll learn what's going on.

Secondly, you learn what it is you love. You definitely won't find it on the outside. That's part of my immigrant mentality coming through. I came here 10 years ago and sort of just went for stuff, but I still believe it's a factor, so I just go for it. Knock on someone's door and say, give us a job. Get inside the walls. You could walk into most small Internet companies today, hang around the office for the day and have a job by 5:00 because I guarantee people would be running by and giving you things to do. I've seen it happen so many times. I think the guy delivering pizza one day got himself a job just because he was standing by the copy machine. It's like, all right I need 25 copies of that and I need them over there by tomorrow. I think he just did it; true story, and they just gave him a job.

I've seen a lot of educational courses that people pitch to these kids. You know, become an Internet superstar and in the end it costs them a lot of money. I'm not against that. If that's what you want, go for it, but I think if you really believe you want to do it and depending on your skill level, you should go in and see how you can get into the organization. From the inside, you can learn what you want to do. Maybe you want to go into marketing, maybe you don't want to go into technology. Because there is such a young, smart bunch of people in this industry, they tend to share a lot more than you would find in a financial institution. Those people are very guarded about their knowledge and sharing and nobody at so-and-so bank wants to tell you anything because that undermines their position. However, I found the opposite in these companies. Because they're young, they do want to share.

Becoming a Leader

I think having the ability to respect people is a big asset for anyone in a leadership position. I've probably interviewed 500 people over the last couple of years and there's one common factor about most of them. They have two issues, not really about money oddly enough. Some of them do, but mostly they can make money if they're good at all. The two issues are basically respect and learning. The standard story is, I'm not getting any respect where I am, my boss is a jerk and I'm not learning anything. That tells you something. That's real time market research.

One of the most important things is if you're getting paid $4 a week, then that's an issue. If you're assuming that you're somehow financially taken care of, then what's your biggest concern? It's really how people treat you and it's how your boss treats you. A lot of times if you get treated with respect and if you're learning, then you'll probably stay there even if the company's gone off the side of a cliff because you have loyalty and you know people have gone to bat for you. I focus on trying to let the people that work for me learn as much as they can because it keeps them interested, it improves them, it makes them better resources for me and in a way both parties win. The company wins because they're getting more talented and smart people, and the employee wins because they're getting the opportunity to learn and feel fulfilled about what they do. It's a simple matter of having respect for people. Just understand that people are not just numbers, people are individuals who work hard.

To make people happy it can be simple things like just saying hello. The CEO says hello to people in the corridor or buys them a drink every now and then. Simple common sense stuff. A lot of the stuff that we do, you know you talk to some people in this industry and it's like they invented everything. My grandfather would probably know more than they'll ever know and he just basically worked lifting boxes off a ship. Basically he did know some things and he knew that if you treat people with respect, it means a lot more than most other things. It means more than any other degree you can get and any master's course anywhere. Help people learn and treat them with respect and always show you know what you're doing even if you don't.

"Becoming a leader" is somewhat oxymoronic, as many would argue the core qualities associated with being a leader are innate not learned. I think the truth may be in the middle somewhere – some qualities we learn and have to work at, and some come natural.

First and foremost, I think empathy and understanding is an essential quality, determination, self-awareness, a sense of humor, passion, perspicacity and a willingness to work hard are all up there too. To become a leader you must possess some of these qualities and be willing to work at the rest. And I think ultimately you have to like and respect the people you lead and believe in what you do. If you have all of the above you may well become a leader.

The Dream Technology

I think it would eradicate nonsense. I don't know, you could probably set it off in a park somewhere and maybe it would beam and eradicate all the nonsense in the area. I have an interesting philosophy on life. I got it from a bumper sticker from a surfer guy at the beach one day in front of me and it said: Life is short, don't be a dick. I pretty much thought that's my philosophy on life. You can pretty much play that to most situations. I think my answer would be global nonsense eradicator. You could point it at Saddam Hussein and a couple of presidents and some marketing executives and a couple of liars and a couple of CEO's and people like that and get rid of the nonsense. People could actually say what they feel and be honest with each other.

Dermot McCormack is Founding Technologist and Chief Technology Officer of Flooz.com. Dermot is charged with developing and implementing all of Flooz.com's technology and information systems and providing strategic direction. Prior to founding Flooz.com, he was Director of Technology for Sponsorship Sales at iVillage where he managed all advertising-related technologies and set up systems that served over 100 million ads monthly. He also developed large-scale Web projects for clients including Sony, Microsoft, and Intel. Dermot is a seasoned technologist who has been involved in launching over 200 Internet sites since 1993. Prior to iVillage, he was the Chief Technology Officer and co-founder of Inw@re Technologies, which he eventually sold in 1996. Dermot received degrees in Electronic Engineering from Dublin Institute of Technology, Kevin St and Trinity College Dublin.

PAVAN NIGAM

Building a Simple & Scaleable Technology Interface

WebMD

Former CTO, Co-Founder of Healtheon

Role as CTO

Well, Jim and I started the whole thing four years ago and right from the first day my role was to get the services implemented. My job included being focused on technology as well as development responsibilities. Along with that, I manage all our Internet product developments, all our divisions, and any Internet base services, including running them from the operational side. So that's about 1500 or 2000 people that report to me.

The Road to WebMD

Let me give you some quick background. I started my career at Intel back in 1982 soon after graduating from the University of Wisconsin at Madison. When I joined the company, Intel 8086 was just out, people were working on the software for the 286. Intel at that time was a hardware company, so we were creating a support environment for developers for the 286. When I exited the company the 386 was getting ready to be released.

After that experience I was recruited by Silicon Graphics to start a full team to create a very sophisticated development environment for software developers for SGI Machines. I created a team there, and the first product released a year after I joined became the best-selling software product at SGI and it maintained that spot for a long time. That's what I did during my first three years at Silicon Graphics.

In my last two years at SGI, I got involved with interactive television. This was back in 1993, when interactive television was going to solve all of the world's problems and the biggest deal was the one between Time-Warner and Silicon Graphics to develop the world's "fanciest" interactive television system. Time-Warner was going to do this pilot in Orlando, Florida, for interactive television. SGI signed the deal and my good buddy Jim Clark called me up and said okay, now you've got to go and execute. I had one year to build a team to create all kinds of new technologies and deploy it. It appeared to be and in fact ended up being one of the most highly publicized interactive television trials.

I spent a year doing that, and that was the most intense and energizing year I've had in my career. It also involved some of the smartest people. We had to create all kinds of new technology in terms of media storage, media delivery, media management, everything. In '94 there weren't very many sophisticated media-based tools around, and we had to deploy it. From the technical perspective, that deployment went fantastic. We had the whole world's media there and we got a lot of publicity. However, from a commercial point of view the whole thing was a disaster because the economics didn't make any sense. When I was leading the whole effort I had a team of around 150 engineers at SGI.

Areas of Expertise

I would say building platforms and applications for mission critical services are my areas of expertise. Mission critical was the buzzword when we were working on interactive television. When you're watching a movie, for instance, if a glitch occurs or a hard disc fails or servers go down, the end consumer can't notice anything. That means your system had to be smart enough to react to all the breakages and had to be scalable and secure. What if suddenly 400 people all at once wanted to watch the same movie—the system needs to be able to react to that. We spent a lot of time building it.

When we started Healtheon four years ago one of the first things we did was say that health care is a field which is as mission critical as it gets, probably even more so than the financial services industry. A breach of anything—of security, of scalability, of anything—could be disastrous for this whole thing. We spent the first year at the company just building the mission critical platform for Healtheon which is still there, it's holding up. We built a lot of services on top of it, so if you're looking at core areas of expertise it's building a platform and then application services on top of it which are targeted toward a vertical market, which is mission critical in nature.

New Technology on the Horizon

Well, by and large the most intellectual discussions we had were on the wireless and broadband fronts. Now it appears that we are coming to a point where all that stuff might become real because all the integrated technologies and products are on the horizon. Broadband is important in the health care field because that can allow a kind of tele-medicine concept, where people can actually go in and interact with each other over an electronic medium.

Wireless is important because today health data is so fragmented and stored in so many databases that it's tough enough to be inside the wired world and access all the data. Wireless allows you to go in and unlock so many of these databases. One scenario is that a doctor could be on vacation and a patient needs a recommendation. The doctor, using the Palm VII device can go and quickly access that patient's information and make a recommendation via wireless email. The doctor can look at things like drug interaction, that type of thing. Wireless is a big deal for us, and I think that it's going to open up services in health care that were just not possible in the wired world.

There are many up and coming things, but those are the two that kind of stand out. And let me add a third one which is important: an Internet security appliance, like a firewall. These are things which you can put in on a small-scale site, like a doctor's office, and unlock the databases. So much of health care data is stored in small databases, whereas the focus so far has been on insurance companies. These big companies have the infrastructure to go and support whatever's needed, but the small sites like doctor's offices cannot afford it.

Implementing Technology in the Health Care Industry

Health care is a very conservative industry so people take longer to embrace any new technology. However, what we're essentially finding is that health care, while being so far behind in technologies, is becoming energetic in adopting some of these newer technologies to the point where it's sort of leapfrogging a middle step. Health care never went through the middle state at all, it's jumping from the mainframe age straight to the Internet age.

In general, though, it is conservative and compared to something like financial services this is an industry that tends to move a lot slower. To some degree, if it were only WebMD pushing people it would have been a tougher job. The good news in all this stuff is that it's a new age and there's pressure on the CEOs and CIOs of companies to go and do something fast because they have the fear of falling behind in the Internet age. To some degree the whole revolution that's happening around us is working in our favor, so we have less pushing to do.

Developing a Strategy to Deal With Changes

Well, you have to do two things. One is that we do have a product-oriented crew in the company. We want everything to be driven by the services we provide, not driven by the technology—this is unlike some other fields. This is unlike my own personal background before WebMD and before Healtheon; I was at Intel and Silicon Graphics where everything was so technology driven. At WebMD we are very much services driven, so the questions are what do the services require, what technologies are available for it and what technologies do we need? To some degree our focus has always been on the quality of the services we need to offer.

Having said that, we do look at the different technologies. As you can imagine, 90% of what you read nowadays is all marketing, so one of the big jobs I and my senior architects have is filtering through that stuff and finding out what's real and what's just hype. It's difficult but I think that's where being in Silicon Valley helps a lot, because you won't believe how much of the stuff we discover through personal contacts and people we've known at various companies rather than going to marketing literature. I don't go and read the latest Oracle literature to find out what's available, I would rather go ahead and call some of my buddies at Oracle or I would go and call some of my buddies who are using stuff from there. I can ask them, okay, what is real and what is not? That's how we get our real answers.

Managing Technical Staff

There are two hats I wear and therefore two organizations that I manage. One hat is a pure CTO hat, and for that I have a CTO architecture council that has our senior brains in it. Within WebMD we have a hierarchy of fellows and chief engineers and principal engineers. In a company of 7,000 today, we cumulatively have less than a dozen of them, so you can imagine what a brain trust that is. That is what I call the CTO architecture council, and that is the group I'm looking to in terms of staying on top of all the latest technology developments.

The other hat I wear is in my role as executive vice president of all our product divisions, all our Internet product divisions—I have five of them today. The five divisions all have general managers who hold responsibility. One is a consumer division which is focused on Internet consumer services; the second is a provider division which is focused on the services we provide to doctors; the third is an employer division, which is services we provide to employer Internet services; the fourth division is what we call the platform services, and it is all the core underlying technologies we develop for supporting these and user services; and the fifth division is more the network operation side of it, managing the operation side of all these different services. All five divisions are reporting to me as executive vice president of all the product services. The overall size of the team as I mentioned before is somewhere between 1500 and 2000, so it's a fairly large size.

Managing is easy for me because I have very strong general managers for all these five divisions. They have to be held responsible, so they go ahead and do what's needed for them. The challenge WebMD has today is to keep track of the latest technology and services, and also to make sure that we are exploiting facilities between the various services and technologies. That's where the CTO architecture council makes recommendations about how we should be using wireless technologies in a consistent way across everything we do at WebMD, or that we should be declaring XML as a standard for everything across what's going on, etc. We don't want every division to have its own technology strategy, so we've created this thing on the side with some of the best brains. These brains belong to the various divisions also, so we keep it very non-threatening in nature. That's how I manage it. So you have good general managers and you have some of the smartest technology brains and you make them all stay marching in the same direction.

Integrating Different Technologies With Acquisitions

Well, as you can imagine it's not easy, but we do two things. First, for some of the fundamental technologies, like what platform the portal is going to run on, we make the decision right away and we use the mission critical platform that Healtheon developed right in the beginning as the fundamental platform for everything. When we acquire a new company that thing goes with the territory. We have to go ahead and use the one platform as the delivery platform for all the services. So we don't spend time arguing every time we make an acquisition.

However, the second thing is that we believe in staging different technologies, so just because we've acquired a company, we don't require the company to go and transform everything to some new set of technologies. If what they have works fine, then we integrate that into our system and we evolve. Then we can take a look at what makes sense and what doesn't make sense, do the portal, do the core platform technologies, etc. Many times you don't have to go and make anything, you could treat one of our acquisitions like a third-party service or a third-party technology base—all you do is integrate into your portal. Let's take Yahoo as an example: The core platform is whatever Yahoo has, but it also has so many services and virtually every service Yahoo has is delivered by third parties.

In our case, if you go to our consumer portal, the core technology is one singular WebMD technology. How do you do a single sign-on, how is security managed, how is personalization managed, how is navigation managed? Everything is done in one and only one way. However, the actual delivery of the service—which might be entering a lab order or entering a pharmacy order—can be carried out by different technologies. If it works, then in the beginning we just integrate it like any third-party service and we punt on the decision as to what level of technology conversion we want to do. It becomes more of a business decision about what is going to cost us more. It might cost us $2 million to port something and the benefit might not be very substantial.

So this is how we can do it. When I explain this to Wall Street I just say that we're pragmatic about it. The core portal will have to be one and only one, and has to be an experience you're not going to compromise on; however, the delivery of the actual services at the back end, whether on a Solaris platform or a Microsoft platform or whatever, is something where we'll make a change when and if it makes sense.

Being Able to Harness Huge Amounts of Data

You have to make sure your overall architecture is very modular in nature. If you're looking from the left to the right, just imagine that the first vertical block talks about the various application services, then imagine a second vertical block to the right which is the portal framework, then imagine a third vertical block to the right which is platform technologies, imagine a fourth vertical block to the right which talks about switching systems, and finally the fifth and last vertical block to the right is executions, sending information to the labs and the insurance companies. Just imagine at the top of the middle one where you had platform technologies that you draw another circle, which is WebMD services, and below the platform block you can draw something that says third-party services. What you have in front of you is our architecture.

What we've done is keep the interface between every vertical block pretty clean, so when you interface from a platform technology to the switching systems you find a very clean interface there for any switch to integrate into our platform. Now, when you have switches you might have gotten from, say, an envoy acquisition that we made recently, all we do is tell each one of the switches to just conform through the simple interface we have to be able to talk to a platform. When we deal with the decision of which switches to sunset and which switches are going to stay on for the long haul, we can just make the decision independent of the platform. We can decide whether we're going to hone in on the envoy site and sunset some of these other switches. That's a decision that can be made inside the vertical, and whichever way we resolve it, it isn't going to impact the rest of the architecture diagram.

Similarly if you go to the last vertical, the execution side, you might discover that United Health Care is there and you might see that United Health Care today has transactions coming from three different switches, but there's only one interface. We can decide which two links to sunset and which is the one we're going to be preserving for the long haul. You can move to the left and the same kind of logic applies to all the application services you have—it could be something like data entry, it could be something like tele-medicine, it could be something like medical transcription. All we're saying is that there is only one way they can talk to the portal and the way to talk is very well defined, and has a security guideline, a personalization guideline, a navigation guideline, and so on.

So you can keep adding more services as long as they conform to the portal framework. Everything works together and we can decide what services to add and to keep and so on. We have to keep our architecture very modular. That's the only way you can keep it manageable, because otherwise it's very easy to lose control over a situation that's very complex.

Developing a Secure Framework

Security was so fundamental to us from day one of the company. People talk less about security now, believe it or not, than they use to four years ago. There were all these headlines all the time about security on the Internet and so on, so we took it very seriously. There are multiple aspects of security, starting with authentication of the individual. Of course for that we are big users of digital certificates, and even do some of that ourselves. We accept certificates from a variety of leading people, for instance our systems use VeriSign to announce our own vendors.

For communication of the transaction and data to our main servers we use SSL, and again we use 128-bit where security is paramount—mainly for aggregated data. We use 40 for less sensitive data, and where the concern is more that we need the users to have access on 40-bit browsers and so on. Employer services are mainly on the 40-bit channel, but SSL is certainly something we use for virtually all our transactions.

For the actual storage of the data in the system we have developed a very sophisticated security design, which has a very sophisticated access control mechanism. With this design, services can inherit access rights from the parent services—they can be virtual maps to the real world where doctors might be going and delegating stuff to some other doctor, and clerks might be delegating stuff. So it can react to the real world scenario. Again, it's a very sophisticated kind of access control mechanism to get to what's stored on our servers.

Finally, of course, you have the physical security. I would invite you to come over to our center in Santa Clara and you'll see how much we have on the physical security front. There are multiple tiers of access to different parts of our network operation center, and it's all biometric. You have to use your thumbprint to get in, and there are only a handful of people who can get into the place where the servers are stored and all that kind of stuff. So, we took security very, very seriously.

Advice for Internet Startups

If you're an Internet startup the big thing you have to do in the early stage is to establish all the core technology you will build your stuff on. One has to be very, very careful in terms of what technology you select, which ones you are betting on succeeding and which ones are proven. That's the risk management you have to do. If you bet on too many new technologies, then you'll end up being a tool for the vendors rather than being an Internet company of your own. You obviously have to bet on a few if you want to be in the lead, though, so I think that's a very fundamental decision somebody has to make.

Getting a Startup up and Running

Well, every startup is different so there's no general answer to it. In the early stage of the company your CEO typically ends up being the sales force. Especially being a Silicon Valley startup, and being in the technology domain, the first thing you want to do is have a very solid engineering team and the critical mass of domain experts—people who know enough about customers you're aiming for so that you end up designing the right stuff. In the early stage I would invest my time in product development, both on the engineering side and the product management or marketing side of it. You need to go and get that thing nailed and have it right. For a startup, early sales, early PR, early anything and everything—it's all being done by the senior executives anyway.

Proving the Idea

What we did and what many companies do is pick a few strategic customers and tie the image of your product with them. When we created the first product, Blue Cross Blue Shield of Massachusetts was our first customer and they were the ones who were contributing the money. They were our first customer, we were working so closely with them that the first release was not going to get made unless the customer was using it. You have to tie up with the customer early on and just work on it till it works—that's your proof of concept.

Raising Capital

It varies so much from technology to technology. I think if you look at Silicon Valley today the first thing any investor is looking at is the management talent behind it. Ideas actually become secondary in nature, so if you're going from that point of view those guys care less about what you're doing and more about who the senior-most people are.

In terms of technology there are two things. One, of course, is the fad of the day—today the fad ends up being optical networking equipment and wireless stuff and so on. Right now all the investors are flocking to it, but who knows where they're going to be six months from now.

I think if you're a technology company you want to look at people who are in a similar space making technology that can replace something which exists. Then you want to see which investors backed the people you are trying to replace. If there's a new form of database then you go and look at the previous relation databases, and you see who the venture capitalists were that backed it. What ends up happening is that inside the venture company you'll have the few venture people who did actively invest in those older companies and know their stuff. Because technology is very difficult to explain, you go in for a trained audience over there.

Biggest Mistakes

Well, it's the oldest mistake in the book: They get so infatuated and obsessed with their own technology that they forget the why of it and who's going to use it. At the end of the day some customer has to get some financial benefit out of it. People won't just fund technology because it's sexy, it has to be useful; I think people very often give sexiness a little more priority than actual usefulness.

U.S. Companies in Relation to International Companies

Today the U.S. is dominating the Internet space so, short of wireless, in the Internet space the U.S. is way ahead of everyone. Wireless is the only place where there are some

European companies who have made a lot of good progress.

I don't know how you would measure it in years, because measuring in years means that eventually in Europe and Japan and so on there will be people who will start developing the same technologies. That doesn't happen, the U.S. technologies do become the standards, it's not that suddenly Europe has an Oracle—that's what the standard throughout the world is. I think the U.S. is just increasing its lead on this thing, and I think the Internet era is going to be very U.S. driven.

The Dream Technology

You know, I think it's a very fundamental thing. I think that for technology to really proliferate in the masses it has to be very, very simple. It might be really complicated underneath the covers and everything, but for the mass audience it has to be simple. I think some of these technologies that are coming out that bring all these net appliances into the wireless space in your home should just be made simpler. I think that if you make that very simple, if you can make it so you turn on the switch and everything configures itself, that's what's really going to make all these appliances at home take off.

Pavan Nigam co-founded Healtheon (now WebMD) in February 1996 along with Jim Clark, founder of Silicon Graphics & Netscape Communications, and has most recently served as its Chief Technology Officer and General Manager of its Internet Operations. In his General Management role, he was responsible for Healtheon's entire suite of Internet-based ehealth services. Recently, he announced his retirement from the Company and is currently playing a Strategic Advisor role. During his tenure, he led the company's transformation from a concept to a company approaching a billion dollars in annual revenues, over six thousand employees and several billions in market capitalization. Michael Lewis extensively profiled his role in last year's bestseller "The New New Thing."

Prior to joining the Company, Pavan worked at Silicon Graphics from August 1989 to January 1996, where he was the General Manager for Silicon Graphic's Interactive Media Group and was responsible for deploying Time Warner, Inc.'s Interactive TV project in Orlando, Florida. Prior to 1989, Pavan was employed by Intel Corporation where he led several microprocessor software teams. He holds a B.S.E.E. from the Indian Institute of Technology (Kanpur) and an M.S.C.S. from the University of Wisconsin-Madison.

MICHAEL WOLFE

Designing the Right Technology Solution

Kana Communications

VP, Engineering

Background

I grew up in the military. My father was a Colonel in the Army and we traveled all over the country. I went to eight different schools and developed an interest in math and computer science early on. I got my first computer when I was in eighth grade and I really enjoyed programming and games, but I didn't do a lot with computers academically since there really wasn't that much to do at the time. There wasn't any kind of curriculum in high school that you could really work off of, but when I got to college I got into computer science. I started at Stanford in 1986 and I graduated with a BS in computer science in 1990 then stayed for a Master's Degree in 1991. I took a lot of classes at Stanford in computer science and I did a lot of projects as part of my academics, but my most valuable experience was a teaching program at Stanford called Computer Science 198. It was written up in Fortune magazine a couple of years ago as being an incubator for people who go off to bigger and better things in Silicon Valley and for people who end up with a big network of other Stanford grads, who are good at technology but are also interested in communicating technical topics. Many folks who came from that program have started companies and have done a lot of significant things in Silicon Valley. I spent my sophomore year through my fifth year (which was my graduate year), incorporating a lot of computer science in my curriculum as well as a lot of tutoring and teaching of computer science. A big part of that program is actually teaching other students how to teach sections, create assignments and grade papers. That way you'll end up with not only a very good technology background, but you get very good at looking at technology from a common sense point of view.

You also learn how to communicate it, how to apply the technical problems to the audience and how to communicate the problems to the audience. That's a very under rated skill. People are under the impression that Silicon Valley is run by people who are technology experts, but the reality is even people who are CTO's or Vice Presidents of Engineering for companies must have communications skills and common sense skills. They need to be able to apply the right technology to the right problem, which is a much more important skill than just your core ability to write a program or do a design. This is the biggest predictor of success in a leadership role like CEO, CTO or VP of engineering.

When I graduated, the Silicon Valley economy was pretty slow and most technical folks were choosing to work for Oracle, Microsoft, Anderson, or for other consulting firms. There was a lag in the startup economy in the mid-90's. Lots of folks were going to Oracle, but graduating from college and driving across the street to work at Oracle with all your friends who you went to college with is an easy transition and a lot of fun, but it's not a good way to make a transition into adulthood. I instead went to work for Goldman Sachs on Wall Street. At the time, the investment banks had all made the transition from technology as a back office tool to keep track of trades and accounts and to a competitive weapon used for trading and real-time market analysis. I worked in the equity derivatives trading group - which in particular are the biggest weapon that the good investment banks had. The good trading floors had very advanced trading systems. I worked with a lot of rocket scientists, literally rocket scientists, who had PhD's in Math and Physics who worked all kinds of crazy derivative models and some pretty high tech stuff. I was working for a very well run, mature company with a very good culture.

Unfortunately, most companies in Silicon Valley cannot be described like that and even the more successful ones just don't have a professional culture. Oracle is an example of a company that's big and successful, but is a relatively unprofessional environment relative to a company like Goldman Sachs, which is one of the better run companies. At the time though, the kind of technology that the banks were doing was high performance systems, lots of object oriented design and development — and very fast work stations, Goldman Sachs even had a few super computers lying around. Wall Street was where a lot of good technology was happening, so it was really an opportunity to combine a good business experience and mega technology experience. It was a good transition between college and the working world and also between technology as an academic subject and technology as an actual tool, — learning engineering methodology, learning how to take a business problem and map the systems onto the problem, which is still the hardest problem to solve.

I learned a lot of those lessons quickly because I found that Stanford especially had a very academic computer science program, but it didn't have the more pragmatic, professional aspect to it even though Stanford has a very tight connection between the industry and the university. Stanford focuses more on theory vs. good engineering practices, like how to work in groups, quality methodology, release methodology and project management, which are the kinds of things that you might learn in a traditional engineering management curriculum.

Areas of Expertise

My academic background is in computer science with a focus on artificial intelligence. I have done some artificial intelligence in my last couple of positions, but I would say my more valuable expertise is taking a business problem and designing the right technology solution for that problem for that market and then implementing it. This requires some balance between object oriented design skills, database skills, Web skills, and being a good business analyst who can map business problems to technology solutions. It also involves staying current and understanding a good breadth of technologies, understanding that you cannot be deep in more than a few areas at once.

Changes in Technology

My current vantage point comes from developing software to help companies run their business, from dot.coms all of the way to Cisco and GM, but I would have a different point of view if I was running an e-tailing Web site or if I was building financial software. When the Internet happened in 1994-95, people were willing to spend a lot of money on Internet point solutions and products that solved a very specific Internet problem, like how to sell on your Web site or how to conduct customer service or how to help people inside of a company collaborate using intranet technology. Those applications are as compelling or even more compelling now than they've ever been, but people are now also realizing that the business problems they're trying to solve really aren't that different than the business problems they've always been trying to solve, namely designing good products, keeping your customers happy, and running efficiently. Now the Internet gives you some interesting new ways to achieve that. In particular, it puts a lot more power in the customer's hands so the customer has lots of ways they can communicate with the company, lots of touch points, and more ability for the customer to control their relationship. For example, you get help for a problem by going to the web site instead of calling an 800 number. That puts a lot of control in the consumers' hand versus relying on that call center operator to be the buffer between the consumer and the company that they're dealing with.

The transition that's happened in the last couple of years is that people now want more integrated solutions. They want to be able to do business on the Web, but they also want the ability to use the same processes and the same internal administrators whether it's sales reps, call center agents across phone, retail, wireless, email, and chat. The Internet gives you a new set of communication channels. It's given businesses the ability to do operations across more channels. If you look at a dot.com like Cooking.com, the only channel they have is to sell online and the only channel they have to market and service is online, then we're going to find that as the more traditional, larger companies mature and learn how to use the Internet channels, it's not going to take them long to build their Internet businesses. The fundamental advantage they have is in terms of size, the history they have, the brand they have and now the technology is mature enough that it will speed up the process. They can buy software off the shelf from companies like ours that help them build their Web presence faster.

The startups no longer have the advantage they used to have. Suddenly, the multiple communication channels, the ability for a company to sell widgets in catalogs, in retail stores, over the phone and on the Web have become a real advantage. The trend is going to be that the Internet businesses that had a unique business, something that couldn't be replicated in the real world (eBay and Yahoo are good examples of that), are going to continue to be very solid companies. The ones that have a disadvantage over the traditional businesses, who now have understood the Web are the ones that are going to need to do something, they're going to need to be acquired by one of these "traditional" retailers to survive.

Creating Good Experiences for the Consumer

Fortunately, one of the businesses my company is in is helping to solve that problem and it will help a lot of our customers do that. From the point of view of the consumer, the thing that they want to achieve with new technology and new communication channels is probably not that different from what they wanted to achieve all along. They want the products they want. They want good prices, they want convenience, they want service, they want sports, weather, and stock prices. You want the consumers' interaction with the people they do business with to be a continuous interaction whether it's wireless, email, Web, phone or in store. That consumer wants the business to have an idea of who they are. They want the history of their interaction with the company to be tracked. Not to the extent that it's intrusive, but there's nothing worse than sending an email saying that you have a problem and then picking up the phone and talking to a representative who has no idea what you're talking about. Regardless of what technology channel a customer uses to contact you, you want to give them a unified experience. The technology is moving a lot faster than consumers and companies understand what to do with it. When a new technology, like the Web or wireless is introduced, there's a huge inefficiency in the market in terms of the gap between what the technology has enabled you to do vs. how much people have capitalized on that opportunity.

When the Internet happened in 1994, there was suddenly a giant gap between what the technology enabled and what businesses were doing with the technology. That's why companies with billions of dollars of revenue and market capitalizations were created in just a couple of years, like eBay, Yahoo, Priceline and eToys. There was inefficiency in the market: a gap between what was possible and what had been done. With the Web, the size of that gap has shrunken. But coming online now is wireless, broadband, XML, new devices, and there are new gaps emerging that need to be filled. There's now a whole new set of technologies that a whole new set of companies are going to take advantage of. There will be more Yahoos and eBay's someday, but we don't know who they are or when they're going to be there. But they're going to get there by taking advantage of new technologies and probably not by continuing to do more and more with the existing Web technology.

Exciting Technologies on the Brink

The most exciting is wireless. Wireless gives you everything that the Internet does with a lot of extras. Wireless is geographically neutral, so you can be anywhere and have the same experience. It can be more personalized in that your wireless device is usually only being used by you. It's not part of a corporate network. It's not a machine that's shared by other people. It's a very personalized device, and that device can be designed to be as intrusive or not intrusive. You can set it to page you, beep you and even shut it off. You can have it vibrate so you can pick the right level of interruption or interaction. You control that level, where with a PC if you're not sitting on you butt in front of it, then you're disconnected. So you have a lot of very binary decisions to make about being online or off-line at any point in time.

With a wireless device you can do voice, Web, data, messaging or chat 24 hours a day. You can pro-actively begin the communication by sending a message to somebody or going to a web site. Because of broadcast and because of peer to peer communications, there's a whole new set of services that you can give somebody with a wireless device that you can't give them otherwise. I go to a new city and I type in that I'm hungry for dinner. I type in what type of cuisine I want, and I say I want a restaurant that has a reservation at 7:00. A reservation appears with a map to the restaurant. They are probably only a couple of years away from being in the mass market in the U.S., and they are already happening in Japan and Europe. Walking into a store and having that store recognize who you are as a consumer is useful. Giving you a special promotion, remembering that a product that you wanted last time you were in the store is now available in your size and sending you a message that tells you that is very possible. I think there are a lot of scenarios that we haven't thought of.

New Technologies Being Developed

The pace of basic technology change is definitely increasing. There are a lot of technologies like wireless that are moving quickly. When you take a new technology and combine it with the things the old technology gave you, and with all the business ideas that have already been executed using the existing technology, there's almost an exponential explosion in the capabilities and the potential. Instead of being a linear progression, I think it gets more exponential.

Specific Technologies at the Forefront: Java/XML

Java is definitely here to stay. Kana was one of the first to start using Java for an enterprise software application back in 1997. We got questions about scalability and long-term viability of Java. Now we're selling to some of the biggest companies in the world like Cisco and GE. They're looking for vendors who do Java for Internet applications and they are skeptical of the ones that don't. In many cases they're developing a lot of internal systems and they rely on Java. Java can keep improving, but it transitioned very quickly from being a leading edge technology to being very mainstream. I think it's definitely here to stay.

Another interesting new technology is SOAP, (Standard Object Access Protocol). It is a way to use XML to let business services on the Internet talk to each other and share data. The Internet is used right now to share information for people to take a look at. XML was designed to solve a problem of moving data around the Internet between systems and not just people. SOAP is a way to use XML to make requests of business processes on other nodes all over the network, which includes wireless devices, laptops, and mainframes. Any computing device can use SOAP.

The Changing Role of the CTO

The role of the CTO is more of a business leader and communicator, as well as being a technical visionary, which is the more traditional definition. The visionary piece of it is understanding the business that the company is in and understanding the available technologies as well as the technologies that are right around the corner, and using all that information to create a technology strategy for the company. The CTO is an increasingly outward looking person who needs to be able to talk to clients, speak at trade shows and talk to analysts. They also need to explain not only the technology the company has, but also try to project why the company has some fundamental competitive advantages because it's using technology or has a knowledge of certain technologies that other companies just don't have.

Keeping an Edge

Keeping up can be overwhelming. The ability to work hands-on with every new technology to get an intimate understanding of it is going down since there are so many to learn. The best way, speaking for myself, is going to the experts. I'll sit down with a developer who is working on a piece of our system and ask them to walk me through what we're doing and what they would like to do next. Visiting with customers, prospects and companies that we might have partnerships with is important as well. It's also important to listen to what they're doing with technology and where they're going. There's a constant diet of material that you have to read: industry magazines, the Web, books. For every new technology, fifty books are published, so you need to find the one or two that will give you what you need. It does require constant study and constant learning.

Staying on top of Technology

You have to strike a balance between depth and breadth. When you're in an academic environment depth is rewarded. It's easy to graduate from college with deep technical knowledge, but not really have much of an idea how those things translate into real world problems. Unless you want to go into teaching or research, which not many people do, you need to start finding business problems that need solving. It's almost like a doctor who knows how to do an operation, but doesn't know why smoking is a bad idea. You need to know the larger context with which you are working. To make technology their career, you need to go deep into some of the fundamentals, but you also need to keep up with the trade press, network, and go to different companies and understand how they position their technology. You need to look more at how people are using technology to execute their strategy, their business strategy. I would probably say breadth over depth in general. I think the more senior somebody gets the more they need to focus on breadth. Early on it's okay to be a little more deep.

Where to get Started

My knowledge is very much around software and computer science. I am not a hardware person, I'm not a biotech person, I'm not a networking person, so it's really hard for me to say. So I see a lot of the opportunities in software. Software drives all of these devices, applications, communication channels, and permeates everything we do. If you look at wireless and the different Internet standards and protocols that are being developed, you'll see that software crosses boundaries. Software is what makes wireless devices work. Software is what makes Web sites work. Software is what makes internal corporate operations happen, so software has become the glue that holds a lot of technologies together.

Skills in Demand

Right now the technologies that are high in demand are Java, object oriented design and development experience, web development, knowing how to build Web pages, how to write and knowing how to actually program Web sites are good skills. It's more than just presenting content to a user.

Another important skill is with respect to databases. Databases are still the foundation for pretty much all software at this point. Most applications are still very much data-driven. You need to be grounded in these base technologies, but I think the more important skill set to have is good engineering expertise and knowing how to actually translate an idea into a project and executing the project.

Being a good communicator and working with teams are needed because the interpersonal issues, project management and business skills are the ones that make those experiences important. Those are the ones that make those technical skills important. You have to be strong in both, and again the more senior somebody wants to get the more they need to be able to leverage their technology skills with communication and business skills.

Changes in Technology From an International Standpoint

The rate of adoption in Europe and the Far East, at least for the kind of Web-based products we build, is about a year to a year and a half behind the U.S. The kinds of problems people are asking us to solve overseas are a bit behind the problems they are asking us to solve in the U.S., both for Web-only businesses as well as the brick and mortars. With wireless, the opposite is true. Our most interesting wireless applications are being driven by our European customers. Japan also has tremendous momentum with wireless data. We are also selling to businesses who want to deploy the same products across the globe, so we are selling to companies who in turn need to roll out capabilities to consumers in many different markets, so we are helping them get there.

Advice for Other Companies Focused on Technology

People often confuse the idea of "product" with "technology." Companies that focus on product tend to succeed. Companies that focus on technologies tend not to succeed. Creating a product is mapping what a customer needs to what you can build. A technology is trying to find customers for something you've already built. We talk to a lot of companies because we do a lot of mergers and acquisitions and partnerships. Most of the companies we talk to are founded as technology companies. A PhD writes an interesting dissertation, and they decide they're going to start a company based on that technology, then they go out searching for clients. There are some counterexamples, but no company has succeeded without quickly becoming customer focused. Companies that create excellent technology to solve real problems win. Sometimes a company like Microsoft is accused with winning only because they have "better marketing," when the reality is usually that they have better marketing, sales, and communicate better, and ultimately build what people need.

Becoming a Leader

The number one skill is communication skills. Learning how to listen, how to understand what people want (employees, customers, partners), and how to communicate to them a vision, a perspective and a plan are very important. Lots of people have good thoughts, but if you can't communicate with them the thoughts don't help anyone. Number two is having the knowledge, intuition, and credibility to make hard choices and help them stick. Sometimes people think that a leader is the one who knows the most. Usually a leader is someone who listens the best and understands their people the best. You have to engage people's emotions, wants and desires and show them a path.

The Dream Technology

If I knew, I would start a company to do it. What I'd like now is the "ultimate device." We have cell phones. We have hand held devices. We have laptops. We have cable TV. We don't have the ability to meld these channels and experiences into one multi-media, multi-channel, high bandwidth device. That device would let you go anywhere in the world, videoconference with anybody, listen to music, watch a movie, watch the news, read email, surf the Web, chat, all with no wires and for free. That'll be the day when I never have to go home again. I would wander the streets completely plugged into the rest of the globe. I do think we're probably less than five years away from that by the way.

Michael Wolfe joined Kana in May 1997 as Director of Engineering and has served as Vice President, Engineering since April 1998. Prior to joining Kana, he served as Director of Engineering at Internet Profiles Corporation, an Internet marketing company. From 1994 to 1995, he was an associate at Wells Fargo Nikko, specializing in software development. Prior to that, Michael was a software-programming analyst at Goldman, Sachs & Co. He has taught computer science at Stanford University and the University of California at Berkeley. Michael holds a B.S. and M.S. in Computer Science from Stanford University.

DANIEL JAYE

The Role of a CTO

Engage

CTO & Co-Founder

Background

I was Director of high performance computing at Fidelity Investments when I came over to start Engage with Dave Wetherell, the chairman of CMGI. My background was in applying parallel database technology to solve customers facing problems, starting out at Anderson Consulting; then I moved on to Epsilon, which was a significant player in the outsource database marketing services area; then I headed over to Fidelity to head up their retail data warehousing efforts.

The Role of a CTO

The role of CTO very much depends on the life stage of the company. In the early days, my role ranged from writing business plans to doing product requirements to writing code to designing software and eventually managing engineering. When you're a very small startup you end up doing everything.

As the organization grew, my efforts focused more and more on the technology and the future of the technology. I was keeping an eye on the business problems for which our company was providing solutions, and insuring that we delivered our technology solutions to our customers on time and with the right future set.

As we grew even larger, eventually my role was separated from tactical responsibilities for delivering specific products and became more focused on insuring that we maintain a leading edge in terms of our technologies, which are still very tightly coupled to the business problems we have to solve. At this point I was guiding the company's business and technology strategies in a linked way. This also involves dealing with a lot of outward facing activities, including acting as the spokesperson of the company, articulating the company's value proposition to investors both in the pre and post-IPO days, as well as representing the company to other external parties such as suppliers and regulators.

How the CTO's Role Differs Across Companies at Similar Stages of Maturity

A lot of it depends on the talent pool of the executive team. If the team has very strong engineering management the CTO doesn't necessarily have to be as involved with guiding the specific delivery of technologies, and can focus more on insuring that the technology solutions are best of breed, the highest quality, innovative, and customer focused. In other companies it may be appropriate given the skill set of the management team for the CTO to have a more operational role. Typically, though, the CTO of a mature company is somewhat separated from that operational role, and is more focused on insuring that the company is delivering the best quality technology solutions in the market—solutions that are going to take the company where it needs to be in the next year or two.

Qualities a CTO Must Possess

There are three top qualities. One is the ability to communicate. The second is the ability to bring together technology solutions and see how they can be combined in an innovative way to deliver a compelling and novel solution. This is a special type of creativity, it's not necessarily inventing things from scratch; very often it's the ability to take different components of a solution and figure out how to combine them in a way to create a novel and valuable solution. And I think the third area is the ability to relate the market requirements to the technology solutions and be able to translate back and forth between those two domains in a very fast manner.

Daily Interactions

On a daily basis, I may be communicating with elected and appointed officials of the government, and foreign governments. I may be dealing with both the marketing and technical executives of our customers, and with marketing and technical executives of partners and vendors. I'll be communicating with the press, both with industry media and general media. I'll be communicating with members of trade and standards associations. I'll be communicating with all the different functions within the company—marketing, sales, engineering, operations—very often in an advisory capacity.

Venture Capitalist Firms and Other Funding Sources

Because we were incubated within CMGI, we had a stable primary funding mechanism for the first few years. That meant that we needed to closely coordinate with the CMGI's strategy. They are unique because they are not just a collection of investments, CMGI is a conglomeration of companies that have been planned very strategically to address critical needs in the marketplace.

In our case, that meant that for almost the first four years here I had an unofficial role as CTO of CMGI, our parent and investor. That had its benefits in that I could coordinate technology summits for an exchange of ideas and partnerships between different CMGI investments, so it created a lot of exposure for Engage within the CMGI community of companies and also let us stay very much in tune with what's happening in the industry. For example, one of the reasons why we bought Accipiter is that there were two CMGI affiliates of ours who had been using their products and loved them. The system is great for references, etc.

I do think that it's critical for the CTO to help develop and maintain relationships with investors because in our market space the technology foundation of the company is critical.

That happened in a special way because we were incubated within CMGI.

The Effect of Incubation

In a very positive way it has brought us tremendous exposure and entrees into lots of different situations, where a small startup by itself might not have received an audience. It also provided an instant network of peers with whom we could very honestly exchange opinions on technologies, vendors, and solutions, and also find out about potentially new competitors, market interests, or new potential companies that we could work with. Once again all the companies that we purchased were companies that we knew of directly or indirectly through our affiliates beforehand. That's all been tremendously beneficial.

I would say that the number one downside I can think of is just the fact that I wish we had done more together earlier. This is not necessarily a negative of being incubated but more that I think in the early years we were all running so fast and in the early days CMGI really didn't have any corporate staff to facilitate synergies between companies. It fell to individuals like myself to try to cause that to happen. I think that we had opportunities to cooperate earlier, but CMGI has started to buildup its capabilities to develop more integrations and synergy.

How the CTO Interacts With the CEO

In my experience it works best when the two of them have a real partnership and a trust. The CEO should trust that the CTO is going to be able to accurately understand and communicate the impacts of technology on business decisions and vice versa, and the CTO should trust that the CEO has the management and decision-making abilities to figure out how to take that advice and apply it to a go-to-market strategy and an operations delivery strategy. There has to be extremely open communication between the two, frequent and open communication to insure that the messages are consistent. They should also act as a sounding board for each other as they fulfill their responsibilities.

The Creation of Engage

When we created the company we knew that the domain we were going to be dealing with, which was Internet database marketing and leveraging data, was going to involve pushing the limits with regard to performance and scalability. We knew that the problems we would be solving would be some of the hardest problems on the planet with regard to performance and scalability and complexity. So the focus was on both integrating and building best of breed technologies for solving those problems and building and insuring that we had an engineering team that had the capabilities to solve those problems.

We started off with the premise that things had to support fundamental principles like partitioning, incremental processing, distributed processing, and extremely efficient interloop processing as we would call it. Those were some of the fundamental principles, but in the end it had to do with making sure we had the right talent pool to build those solutions.

Building a Technology Team

Problem-solving ability is probably first and foremost. Second of all is domain expertise, you want to start off with a core of senior people who have been there, done that before. You want people who are going to be able to work together; there are many very talented people in the industry but sometimes talent comes with an inability to work as part of a team with others. I've always found that in the end you get better results and secondary benefits, such as better morale and a better culture and environment for working when you have people who are not only extremely talented but also balanced individuals who work well as part of a team and with others, and embody principles like mutual respect, mentorship, etc.

Acquiring People

There are a couple of strategic points. One is that we are fortunate enough to have a great pool of people in the company today with these resources, and we do what we can to make sure that they're very happy with what they're doing because they are our greatest asset. I think I have been able to attract top quality talent with those skills in the past, but it really comes down to working with the independent recruiters and with the people directly to really try to focus on people who are athletes. We want people with good strong skills, and are not necessarily after people strictly on the basis of what company they worked for last.

Technologies Engage has Created

There are a few, and I could speak to both the technologies that we've created in Engage from their initial construction, and also to the technologies that we're very proud of that have come into the company through acquisition. The teams that built those technologies have done just a fantastic job in their respective areas.

One of the technologies we were most proud of was the data warehousing technology we built in the early days to solve some of the early problems we faced with scalability and data management. We invented the concept of parallel network data flow processing. Probably one of the things that I regret most is having spent my time building the technology and not taking the time to go off and get a patent on it, but we were heads down trying to get to market. So, there is some technology there that really was and still is very unique on the planet in the area of parallel data base, network data flow processing.

Another area that is core to us today is the way in which we take flows of loosely, temporally sorted clickstream and log data and process it into a picture of a browser's activity at a Web site. Then we apply algorithms that we developed from our parallel data base technology to very efficiently aggregate that data and manage it across a distributed platform at very, very high-end scales. We do it in a way that allows great flexibility in how we deploy the technology, how we customize it, and how we adapt it to many different business requirements. I think that platform for profiling is perhaps our greatest achievement among the technologies that were built organically from within the original Engage group.

Then there are four other technologies that are critical and unique. One is our Ad Manager technology that the original Accipiter team in North Carolina built. That's a group that we bought back in 1998 that we had known for a couple of years, and they had built a great product for allowing publishers to serve and manage their advertising. They've just kept a wonderful focus on improving that technology so that it has maintained a position as the premiere software product for sites to sell and manage their own advertising.

On the AdKnowledge side, which is a group that we acquired at the end of last year, we have the technology that allows advertisers to serve campaigns across a large scale of sites in a very open way. Perhaps the most unique technology there is the analytics capability, where they have the ability to deeply analyze the marketers' activities at a level of detail and with a sophistication that is unrivaled anywhere in the industry. Because we have a background in massive power processing and large-scale data processing in what was originally the Engage group, we have nothing but admiration for what they've been able to accomplish at AdKnowledge.

Another area that we're also huge fans of is the Engage Media organization that came with our acquisitions of Flycast and Adsmart earlier this year. They've built a technology platform for optimizing performance of ad campaigns that is optimizing responses from marketers across, at this point, 4900 Web sites. It is both incredibly fast and incredibly efficient, and once again I think it's unrivaled in the industry, considering both the complexity of the processing that they perform on literally billions of impressions and transactions a month, the efficiency with which they do it, and the response time and through put.

I think the final area that is pretty unique and amazing is the technology platform inside the IPRO Internet auditing and measurement division of Engage. It is designed to handle log file processing for our customers on an outsource basis, and it's the place where we take log files from the largest sites of the Web and deliver thousands of reports each day customized to sites' needs. The reports are generated based on processing billions of transactions for our customers, and there's a distributed data processing facility in that organization that is unique both for the scale complexity and efficiency with which it processes data.

So, in short, we've gone from being a small startup with a focus to being a large company that is providing comprehensive solutions for marketers and publishers, and to being a company that is quite multi-faceted. What we've been able to do is combine best of breed technology solutions in different areas into something that's comprehensive for a marketer and publisher.

The CTO Role in Acquiring Companies

Generally I've been intricately involved in or led the technology due diligence efforts. I've evaluated both the existing technology and the quality of the products. I also look at the makeup, caliber and demeanor of the development groups and the quality of the code they've written, not just how well it's working today, but also the keys to that development team's success in delivering that product and how that can be maintained going forward.

Creating a New Technology Solution From the Bottom Up

I think it starts off with intuition. Successful solutions almost always start with somebody who has experienced the pain of the lack of a solution. In my case I had some understanding from my marketing data warehousing experience of technologies that I needed in order to solve business problems that I just couldn't buy. Usually you need someone, whether it's a product manager, founder, someone in sales and marketing, or a CTO, who has a really good feel and intuition for what's missing in the marketplace. Then what you need on top of that is a vision of what the solution needs to be to solve that problem. That's usually the situation, and it can come from both the bottom up in an organization or it can come from the top down, depending on how the company was started.

Keeping an Edge

You spend a lot of time doing research. I personally find that going to a couple of key conferences helps a lot more than going to conferences all the time. I think that's one of the key benefits of the Internet: You can keep up with a lot of technological innovations through aggressive research on the Internet. We have a lot of strong, high caliber people, so very often a lot of the potential innovations come to my attention because one of the engineers on the team has come across something and posts a message to our internal message board or to me, saying "By the way, have you seen this yet? I think this might be good for solving problem X."

Recommendations on Conferences to Attend

It depends on your domain and subject area. I like to go to the Oracle and Microsoft professional developer conferences, which are the conferences where you actually get the product managers and engineers from those companies attending. Those are usually annual conferences. I like to go to those about once every two years; I don't need to be there every year, but I think once every two years covers enough changes on those platforms and it's a good opportunity.

There are a couple of specific industries that I try to keep up to speed on. I do that by attending the KDD Knowledge Data Management Conference, which is where a lot of the data mining papers are presented. Another one is the VLDB conference, not the commercial one, but the academic one. Those, I tend to try to get to about once every two years. Then I try to keep up in those disciplines in the mean time through a sort of information archaeology, where I'm trying to sift out the information from different web sites, academic papers, and what comes across my wire from people forwarding things to me.

Keeping the Organization Technologically Nimble

One thing is to keep people focused and empowered so you have teams that are tightly knit and focused on delivering specific solutions. Then I periodically try to get the key thought leaders together across the company, because we do development in four different locations within six or seven different development groups. I try to get those people together on a periodic basis so that they can figure out where their place is, and can solve problems together or separately or at least share ideas. The second thing is to make sure there is a rapid internal email dialogue between the different groups, particularly when it comes to problem solving.

Using XML

We're looking at XML in a couple of different ways. One is that our product set is starting to integrate XML in a number of different areas, including allowing integration of some of our components with other technologies. In our service business there is work going on in the industry to try to create XML base standards that will allow us to work with our customers and partners more closely, in a more automated fashion. We're really very driven by the industry in that respect, but we haven't adopted it as aggressively as others just because it hasn't necessarily been the primary business driver. We tend to be customer and market driven, so we build world-class unique technology solutions when we have to, but we don't necessarily go build things just for technology's sake.

General View of XML

XML's a very useful technique for expressing meta-data and exchanging data and information. I don't think it's a panacea by any means, and there's still a tremendous amount of work and overhead required to get a payback on it, but it's going to be as commonplace soon as SQL is today.

The Future of Programming Languages

We tend to focus on the solutions that are mainstream but also extremely effective. For example, most of our critical software is written in C++. The major reason for that is that we have to solve extremely large scale problems and there are very robust and well-proven ways of doing that with C++ today and it performs extremely well when implemented correctly.

One of the major benefits of Java in some environments is that it's so much easier to pick up and use effectively. Its a little bit less dangerous, if you will, than C++. But, as I've said, up till now we've had the talent and skill set to use C++. That being said, pretty much all of our developers prefer Java because it allows them to focus more on the programming and less on some of the housekeeping and trivia development. Where it makes sense we'll embed Java in some of our tools, in some of our administration capabilities, and in areas where performance may be less important because it's a seldom used function, but code development efficiency is important.

Future Dominant Technologies

I think XML is going to become quite dominant from a presentation perspective as people try to find a way to get more re-use out of their presentation layer components. I think XML will end up being a technology that's quite standard for that.

I think that we'll see expanded use of Java, in particular in the form of Java being used to deal with the procedural aspects of the business logic. You'll still continue to see C++ be a primary language for large scale server site applications.

Handling Privacy Issues

That's probably one of the things that makes my job a little bit different than some of the others in the area: We've always used privacy as a critical area for our industry and solutions in our space. I started the company realizing that we need to do business differently on the Web, and that everything we wanted to do, we could do without knowing who individuals were.

In the early days when people first started to say, gee, there might be privacy issues on the Internet, it originally came up in technology circles and people started to propose solutions that were overly broad, that would have inadvertent consequences.

It fell to me to start answering those technology criticisms, providing alternative solutions to solve the legitimate concerns and try not to throw the baby out with the bath water. I started working on specifications at the IETF and was part of the original P3P team to build those solutions and those standards. Over the last few years this debate has become quite public in the media and in the government and there's been a dearth of people in the public policy area that have really had a deep understanding of both the market needs and the technology. In my role as CTO, and partly because of my background in authoring privacy technology standards, it's become necessary for me to spend quite a bit of my time focusing on that.

We believe that privacy is a critical consumer need. I think we were the first to recognize that in our space. We are committed to a solution that protects privacy through anonymity but that still enables a marketing and advertising-driven Internet to flourish. You can have anonymity and still have marketing effectiveness and measurability, and our whole company philosophy and strategy is driven by that premise.

Technology Infringing on Privacy

What I found is that the process has been more of a dialogue where somebody will raise a concern, a privacy implication that somebody may not have fully thought of, and for us at Engage that's always been an opportunity creation situation. Somebody will say, "What about this problem?" And we'll go off and think about solving that problem given our constraints.

For us it's always created opportunities. For example, people start to get concerned about inadvertent leakage of data from one Web site to another; a year and a half ago we started to go back and say, gee, we don't really need this type of data, is there a way in which we could get rid of that data but still make the inferences from it that we need to that aren't privacy sensitive? Because we had a flexible, real time architecture we were able to implement a solution that minimized the amount of data we kept at Engage Knowledge. So when leakages started getting more press—like the recent situation where Microsoft was leaking email addresses and refer URL's—we could then say, that's not actually an issue for us at Engage

Knowledge because we don't have that data. I think, once again, that it's been a challenge for other folks, that some companies do view the privacy needs as an obstacle, but in our business it has created opportunities.

Technology Moving Towards More Privacy

I think we are absolutely moving towards more privacy and in fact, from my perspective in this area, it's an accelerating trend. There was a lot of talk in the past, but now we're seeing real action and implementations. I think it's because mainstream industry is suddenly taking this much more seriously. We've seen some of the standards that were affected years ago now all of a sudden coming to the forefront and we're seeing a lot of traction and momentum—so I think that the technology is rapidly becoming real.

The Future of Cable

We can look at cable from two perspectives: as a broadband ISP and then in terms of convergence. Convergence is going to happen fairly organically. I don't think it's going to happen that one day all of a sudden everybody's bought a box that gives them everything they need. Rather, you'll see that PCs evolve to have more and more of the capabilities that are developing the set top boxes which will evolve to look more like the PCs. The PCs start to develop more of a high end application flavor so that if, for example, you've got digital cameras and you want to play with your digital photos or your home videos you may go for a Compaq, but if you primarily want to do email, draw invitations and communicate and browse, then you might just be fine with a set top box that looks more like what last year's PC looked like, except in a very robust, commoditized package.

In the US, cable will end up being the primary mechanism by which households get broadband connectivity into the broader Internet, with certain exceptions like apartments and other areas. I think there's a place for broadband wireless in the marketplace as well.

In Europe I think it will be quite different, I think in Europe you'll actually see broadband wireless. I think you'll see cable take on a similar evolutionary role to cable in the US at least in places like Germany and the UK, but I think elsewhere in Europe, and in Germany and the UK as well, you'll see ADSL on one side and broadband wireless on another side. Each will probably overtake cable in most countries.

The AOL/Time Warner Merger

I think that they've got to tread very carefully. Programming and distribution have always been an issue in this marketplace; I think that you'll see them work very carefully to try to ensure that while they may have relationships they very carefully keep proper neutrality between connectivity, programming and distribution. They'll do some connectivity, some programming, and some distribution, but they'll make sure that in all those different tiers there are multiple players.

The Future of Wireless

We're going to have broadband wireless, but I don't think it's going to be the primary use. The primary use of wireless is going to continue to be for voice. I think there will be voice applications and we'll have some things related to that. Maybe we'll have voice phone calls potentially subsidized by advertising. You'll have microbrowser applications that start to get very sophisticated as both the screen resolutions and the batteries and the bandwidth improves, and I think you'll have PDA applications where people opt for a larger form factor solution because they're doing more serious PDA activities. I think you'll see a bunch of flavors there, but it's going to remain volatile and evolutionary for the next four or five years.

Broadband Satellite Access to the Internet

Broadband satellite is going to go the way of the dodo in the long term, except in very remote areas where it's just not economical to do the land line carriage like certain areas in South America, Central Asia, and India. Otherwise the infrastructure connected to the coastline is too close and it's just too easy to run a trawler with a big reel of fiber optic cable behind it to any coastline. I know the need for it is going up and up as well, but I'm a big believer in our ingenuity and figuring out ways of getting more bits through the fiber.

Exciting Aspects of Being CTO

The diversity of what I get to experience in a day is the most exciting. It's going to vary day to day, of course, but it's a role where you can be up to your neck in discussions about what types of algorithms you should use to solve a particular problem one day, and the next day be talking with a customer about how you could increase return on investments or certain types of marketing campaigns, and the day after that be talking about the role of technology and regulation and legislation across jurisdictions. I don't know of any other job in the world where you'd have that much variety.

The Dream Technology

We're doing a great job with the soft technologies. If we were not limited to the technologies I personally work on today, I would think about a new revolutionary mechanism of safe, efficient, non-polluting, high-speed transport. I think it's great that we're getting people closer together virtually, but sometimes there's no substitution for being there. I wish there was an easier, faster, cheaper, safer way to get from the West Coast to the East Coast or from the US to India.

As CTO and co-founder of Engage, Daniel Jaye is responsible for guiding technology strategy and corporate research and development as well as establishing Engage's industry leading privacy position. Daniel has pioneered interactive marketing and data warehousing technologies and invented several of Engage's patent-pending profiling and privacy technologies. Daniel has also assisted with technology strategy, planning, and coordination for Engage's majority shareholder, CMGi.

Daniel is actively involved with several online privacy initiatives. He is a founding member and active participant in the Platform for Privacy Project P3P) at the World Wide Web Consortium (W3C), was recently appointed by the Federal Trade Commission to serve on their Advisory Committee on Online Access and Security, and the Customer Profile Exchange standard (CPExchange). In addition, Dan is the author of an Internet Draft submitted to the Internet Engineering Task Force (IETF) that allows Web servers to inform users of their privacy practices with regard to HTTP cookies. Daniel is a recognized expert on internet privacy and a frequent speaker on the topic at industry events, privacy conferences, and workshops hosted by the Department of Commerce and the Federal Trade Commission. This past year, Daniel testified before the full Senate Commerce Committee and also its Science, Technology And Space subcommittee. During my Prior to co-founding Engage, Daniel was the Program Director for high-performance computing at Fidelity Investments, where he managed Fidelity's retail marketing data warehouse and applications and led projects responsible for enterprise-wide retail customer management and sales process re-engineering. Daniel has also managed the delivery of parallel customer database applications at Epsilon and Andersen Consulting. Daniel holds a B.A. in Astronomy, Astrophysics and Physics from Harvard.

ASPATORE TECHNOLOGY REVIEW
Tear Out This Page and Mail or Fax To:

Aspatore Books, PO Box 883, Bedford, MA 01730
Or Fax To (617) 249-1970

Name:

Email:

Shipping Address:

City: State: Zip:

Billing Address:

City: State: Zip:

Phone:

Lock in at the Current Rates Today-Rates Increase Every Year
Please Check the Desired Length Subscription:

1 Year ($1,090) _____   2 Years (Save 10%-$1,962) _____
5 Years (Save 20%-$4,360) _____   10 Years (Save 30%-$7,630) _____
Lifetime Subscription ($24,980) _____

(If mailing in a check you can skip this section but please read fine print below and sign below)
Credit Card Type (Visa & Mastercard & Amex):

Credit Card Number:

Expiration Date:

Signature:

Would you like us to automatically bill your credit card at the end of your subscription so there is no discontinuity in service? (You can still cancel your subscription at any point before the renewal date.) Please circle: Yes   No

*(Please note the billing address much match the address on file with your credit card company exactly)

Terms & Conditions

We shall send a confirmation receipt to your email address. If ordering from Massachusetts, please add 5% sales tax on the order (not including shipping and handling). If ordering from outside of the US, an additional $51.95 per year will be charged for shipping and handling costs. All issues are paperback and will be shipped as soon as they become available. Sorry, no returns or refunds at any point unless automatic billing is selected, at which point you may cancel at any time before your subscription is renewed (no funds shall be returned however for the period currently subscribed to). Issues that are not already published will be shipped upon publication date. Publication dates are subject to delay-please allow 1-2 weeks for delivery of first issue. If a new issue is not coming out for another month, the issue from the previous quarter will be sent for the first issue. For the most up to date information on publication dates and availability please visit www.Aspatore.com.

CMGI, New Enterprise Associates, Bertelsmann Ventures, TA Associates, Kestrel Venture Management, Blue Rock Capital, Novak Biddle Venture Partners, Mid-Atlantic Venture Funds, Safeguard Scientific, Divine interVentures, and Boston Capital Ventures. Learn how some of the best minds behind the Internet revolution value companies, assess business models, and identify opportunities in the marketplace.

Inside the Minds: Leading Advertisers
Industry Leaders Share Their Knowledge on the Future of Building Brands Through Advertising – *Inside the Minds: Leading Advertisers* features CEOs/Presidents from agencies such as Young & Rubicam, Leo Burnett, Ogilvy, Saatchi & Saatchi, Interpublic Group, Valassis, Grey Global Group and FCB Worldwide. These leading advertisers share their knowledge on the future of the advertising industry, the everlasting effects of the Internet and technology, client relationships, compensation, building and sustaining brands, and other important topics.

Inside the Minds: Leading Consultants
Industry Leaders Share Their Knowledge on the Future of the Consulting Profession and Industry - *Inside the Minds: Leading Consultants* features leading CEOs/Managing Partners from some of the world's largest consulting companies. These industry leaders share their knowledge on the future of the consulting industry, being an effective team player, the everlasting effects of the Internet and technology, compensation, managing client relationships, motivating others, teamwork, the future of the consulting profession and other important topics.

Inside the Minds: Leading CEOs
Industry Leaders Share Their Knowledge on Management, Motivating Others, and Profiting in Any Economy - *Inside the Minds: Leading CEOs* features some of the biggest name, proven CEOs in the world. These highly acclaimed CEOs share their knowledge on management, the Internet and technology, client relationships, compensation, motivating others, building and sustaining a profitable business and making a difference at any level within an organization.

Inside the Minds: Internet Marketing
Industry Experts Reveal the Secrets to Marketing, Advertising, and Building a Successful Brand on the Internet - *Inside the Minds: Internet Marketing* features leading marketing VPs from some of the top Internet companies in the world including Buy.com, 24/7 Media, DoubleClick, Guerrilla Marketing, Viant, MicroStrategy, MyPoints.com, WineShopper.com, Advertising.com and eWanted.com. Their experiences, advice, and stories provide an unprecedented look at the various online and offline strategies involved with building a successful brand on the Internet for companies in every industry. Also examined is calculating return on investment, taking an offline brand online, taking an online brand offline, where the future of Internet marketing is heading, and numerous other issues.

Inside the Minds: Internet Bigwigs
Industry Experts Forecast the Future of the Internet Economy - *Inside the Minds: Internet Bigwigs* features a handful of the leading minds of the Internet and technology revolution. These individuals include executives from Excite (Founder), Beenz.com (CEO), Organic (CEO), Agency.com (Founder), Egghead (CEO), Credite Suisse First Boston (Internet Analyst), CIBC (Internet Analyst) and Sandbox.com. Items discussed include killer-apps for the 21st century, the stock market, emerging industries,

international opportunities, and a plethora of other issues affecting anyone with a "vested interest" in the Internet and technology revolution.

Bigwig Briefs: Management & Leadership
Industry Experts Reveal the Secrets How to Get There, Stay There, and Empower Others That Work For You
Bigwig Briefs: Management & Leadership includes knowledge excerpts from some of the leading executives in the business world. These highly acclaimed executives explain how to break into higher ranks of management, how to become invaluable to your company, and how to empower your team to perform to their utmost potential. (102 Pages) $14.95

Bigwig Briefs: Human Resources & Building a Winning Team
Industry Experts Reveal the Secrets to Hiring, Retaining Employees, Fostering Teamwork, and Building Winning Teams of All Sizes
Bigwig Briefs: Human Resources & Building a Winning Team includes knowledge excerpts from some of the leading executives in the business world. These highly acclaimed executives explain the secrets behind hiring the best employees, incentivizing and retaining key employees, building teamwork, maintaining stability, encouraging innovation, and succeeding as a group. (102 Pages) $14.95

Bigwig Briefs: The Golden Rules of the Internet Economy (After the Shakedown)
Industry Experts Reveal the Most Important Aspects From the First Phase of the Internet Economy
Bigwig Briefs: The Golden Rules of the Internet Economy includes knowledge excerpts from some of the leading business executives in the Internet and Technology industries. These highly acclaimed executives explain where the future of the Internet economy is heading, mistakes to avoid for companies of all sizes, and the keys to long term success. (102 Pages) $14.95

Bigwig Briefs: Startups Keys to Success
Industry Experts Reveal the Secrets to Launching a Successful New Venture
Bigwig Briefs: Startups Keys to Success includes knowledge excerpts from some of the leading VCs, CEOs CFOs, CTOs and business executives in every industry. These highly acclaimed executives explain the secrets behind the financial, marketing, business development, legal, and technical aspects of starting a new venture. (102 Pages) $14.95

Bigwig Briefs: Guerrilla Marketing
The Best of Guerrilla Marketing
Best selling author Jay Levinson shares the now world famous principles behind guerrilla marketing, in the first ever "brief" written on the subject. Items discussed include the Principles Behind Guerrilla Marketing, What Makes a Guerrilla, Attacking the Market, Everyone Is a Marketer, Media Matters, Technology and the Guerrilla Marketer, and Dollars and Sense. A must have for any big time marketing executive, small business owner, entrepreneur, marketer, advertiser, or any one interested in the amazing, proven power of guerrilla marketing. (102 Pages) $14.95

Other Best Selling Business Books Include:

Inside the Minds: Leading Accountants
Inside the Minds: Leading Women
Inside the Minds: Leading Deal Makers
Inside the Minds: Leading Wall St. Investors
Inside the Minds: Leading Investment Bankers
Inside the Minds: Internet BizDev
Inside the Minds: Internet CFOs
Inside the Minds: Internet Lawyers
Inside the Minds: The New Health Care Industry
Inside the Minds: The Financial Services Industry
Inside the Minds: The Media Industry
Inside the Minds: The Real Estate Industry
Inside the Minds: The Automotive Industry
Inside the Minds: The Telecommunications Industry
Bigwig Briefs: Term Sheets & Valuations
Bigwig Briefs: Venture Capital
Bigwig Briefs: Become a CEO
Bigwig Briefs: Become a VP of Marketing
Bigwig Briefs: Become a CTO
Bigwig Briefs: Become a VP of BizDev
Bigwig Briefs: Become a CFO
Bigwig Briefs: The Art of Deal Making
Bigwig Briefs: Career Options for Law School Students
Bigwig Briefs: Career Options for MBAs
Bigwig Briefs: Online Advertising
OneHourWiz: Becoming a Techie
OneHourWiz: Stock Options
OneHourWiz: Public Speaking
OneHourWiz: Making Your First Million
OneHourWiz: Internet Freelancing
OneHourWiz: Personal PR & Making a Name For Yourself
OneHourWiz: Landing Your First Job
OneHourWiz: Internet & Technology Careers (After the
Shakedown)

Go to www.Aspatore.com for a Complete List of Titles!

Also from Aspatore Books:

Bigwig Briefs
Condensed Business Intelligence From Industry Insiders

Become a Part of
Bigwig Briefs

**Publish a Knowledge Excerpt on an Upcoming
Topic (50-5,000 words), Submit an Idea to Write an
Entire Bigwig Brief, Become a Reviewer, Post
Comments on the Topics Mentioned, Read
Expanded Excerpts, Free Excerpts
From Upcoming Briefs**

www.BigwigBriefs.com

Bigwig Briefs features condensed business intelligence from
industry insiders and are the best way for business professionals
to stay on top of the most pressing issues. There are two types of
Bigwig Briefs books: the first is a compilation of excerpts from
various executives on a topic, while the other is a book written
solely by one individual on a specific topic. *Bigwig Briefs* is
also the first interactive book series for business professionals
whereby individuals can submit excerpts (50 to 5,000 words) for
upcoming briefs on a topic they are knowledgeable on
(submissions must be accepted by our editorial review
committee and if accepted they receive a free copy of the book)
or submit an idea to write an entire Bigwig Brief (accepted
ideas/manuscripts receive a standard royalty deal). Bigwig
Briefs is revolutionizing the business book market by providing
the highest quality content, written by leading executives, in the
most condensed format possible for business book readers
worldwide.